MW00709468

EXTERIOR RENOVATION AND RESTORATION OF PRIVATE DWELLINGS

Byron W. Maguire

PTR Prentice Hall, Englewood Cliffs, New Jersey 07632

Library of Congress Cataloging-in-Publication Data

MAGUIRE, BYRON W.,
Exterior renovation and restoration of private dwellings / Byron W. Maguire
p. cm.
Includes index.
ISBN 0-13-296914-9
1. Dwellings—Maintenance and repair. 2. Exterior walls—Maintenance and repair. 3. Garden structures—Maintenance and repair. 4. Roofs—Maintenance and repair. I. Title.
TH4817.M337 1993 92-30703
690'.24—dc20 CIP

Editorial/production supervision
and interior design: *Barbara Marttine*
Cover design: *Wanda Lubelska*
Acquisitions editor: *Bernard Goodwin*
Manufacturing buyer: *Mary McCartney*

© 1993 by PTR Prentice-Hall, Inc.
A Simon & Schuster Company
Englewood Cliffs, NJ 07632

The publisher offers discounts on this book when ordered in bulk quantities. For more information, contact:

Corporate Sales Department
PTR Prentice Hall
113 Sylvan Avenue
Englewood Cliffs, NJ 07632

Phone: 201-592-2863
Fax: 201-592-2249

All rights reserved. No part of this book may be reproduced, in any form or by any means, without permission in writing from the publisher.

Printed in the United States of America

10 9 8 7 6 5 4 3 2 1

ISBN 0-13-296914-9

Prentice-Hall International (UK) Limited, *London*
Prentice-Hall of Australia Pty. Limited, *Sydney*
Prentice-Hall Canada, Inc., *Toronto*
Prentice-Hall Hispanoamericana, S.A., *Mexico*
Prentice-Hall of India Private Limited, *New Delhi*
Prentice-Hall of Japan, Inc., *Tokyo*
Simon & Schuster Asia Pte. Ltd., *Singapore*
Editora Prentice Hall do Brasil, Ltda., *Rio de Janeiro*

CONTENTS

PREFACE

This book on the exterior renovation and restoration of private dwellings has been prepared for both the homeowner and the contractor who must work out renovation or restoration problems for the owner's home. Many jobs are seemingly simple and can be performed by the owner with modest skills and knowledge. Other jobs, also seemingly simple, are well beyond the capabilities of the semiskilled homeowner. They need the skill and knowledge of experts to restore the exterior to its original condition.

The key to deciding what to do about a problem and how to ensure that the job is done with quality is contained in this book. The information is written in common language rather than technical jargon. The details are varied and informative and provide a basis for intelligent discussion between a contractor and homeowner. On the other hand, the details may be sufficient for the owner to attempt to perform some jobs. Some of the details include definitions of the problem as seen by the experienced contractor. Following these discussions are alternative solutions that provide the owner with several solutions, which can be performed at different costs.

It is important to understand what constitutes a contract for restoration. Each project contains information pertaining to contracting for the work. *Statements of Work* describe the specifications that form the foundation of a quality contract. Sample contract descriptions and contents are provided. Some are fixed-price types and others are various time and materials types. Reasons for their selection are provided with relation to fairness to both parties.

Scoping the work effort is often a major problem for the owner. Trying to determine how much material and labor will be required, what types of materials and skills will be used to make the restoration, and other concerns are satisfied in the *Materials Assessment* and *Activities Planning Chart* for the number of days to complete each phase of the work sections in each project.

Beyond these details, we give descriptions on how the work will progress, which provide a sense of the necessary interruption to normal living and an understanding of why quality workmanship takes time.

For the contractor, we have provided a structure for understanding how to approach the homeowners who many be unsure and wary. Winning their confidence that your work is first quality and that their satisfaction is paramount can be a demanding undertaking. Many ideas regarding integrity and interaction are expressed in the text.

Being honest with the owner is important. The contractor is in business to provide a quality service and product, but must also make a profit to stay in business. To this end, the owner of this book will understand the need for adding overhead and profit to the costs of direct labor and materials. Sometimes we address these items as variable costs and fixed costs. We know that no profit can be made until all fixed costs for the year have been covered by funds received from completed contracts. We know that variable costs such as insurance and travel must be accounted for every job; yet they are not materials nor are they labor. We have provided the owner clear facts about these requirements so that he or she can understand what a fair price is and how it is established. The contract need not specify these items, but the work-up sheet in the office will clearly indicate them.

There are eight chapters in the book. Each focuses on a specific part of the exterior of the private dwelling. Within each chapter there are specific projects. The Contents provides their titles. These were specifically chosen to be broad in scope so that many aspects could be discussed. However, I felt that, although they needed to be broad for study purposes, the contractor or owner also needed a quick reference listing that could pinpoint a problem. Therefore, I have included an Appendix with these details. The list identifies 185 different problems and directs where in the text to find details corresponding to them.

Chapter 1 covers a wide range of problems with the roof and cornice. These parts of the house take severe beatings from the weather and sun. A multitude of things can go wrong and most of them are explained and solved. Discussions range from damage by hurricanes, to natural weathering from time and climate, to the deteriorating effects of direct sunlight.

Chapter 2 is the first chapter on exterior walls. Here we described variety of problems with siding and shingles. There are many solutions and these are explained.

Chapter 3 is also on exterior wall covering. Now we deal with brick, blocks, and stucco. These are masonry types and we provide ample details on how to solve problems with cracked joints, settling walls, and ruptured or cracked stucco.

In Chapter 4 we spend a lot of time explaining how damaged exterior wall structures are repaired. Both the western platform and balloon framing types are shown. This chapter is important since storms or flying objects frequently damage studs, destroy window framing, and the like. Solutions to these kinds of problems are included.

In Chapter 5 we find solutions to problems with doors and windows. Because millions of homes have windows over 30 years old, we provide details on how to make repairs to double-hung windows with sash weights. We also discuss how to insulate windows to keep utility cost under control.

Doors are also a problem over time. Every imaginable problem that can go wrong is covered in this chapter. We even provide details on how to make the door fit better and close snugly.

Chapter 6 shifts from the parts that make up the basic house to the decks and porches attached to the house. We know that these items take a severe beating over time. Repairs and restoration are realities that must be done. A wide variety of problems is identified and alternative solutions are provided. These include restoring the quality and serviceability of the perimeter seating, repairing the sagging or broken stairs, and others.

Chapter 7 continues with items attached to the house in a peripheral way. These include damaged sidewalks and driveways and patio slabs made from concrete. Here we discuss the need to improve the durability of the slab or sidewalk to make it last longer than the original.

Chapter 8 focuses on the screened-in porch. Several types of construction are used to create a screened porch. We provide descriptions of each and also identify the problems that can occur. Details on correcting the problems are clear. We also discuss in a project how to solve problems with electrical wiring and fixtures that are giving trouble. Finally, we provide ample detail in correcting problems with posts and columns.

Problems, suggested solutions, materials and labor details, contract definitions and descriptions and studies are given in the sequence that work must be done. This is a handy book to have in your reference file in your home or office. You, the homeowner, can readily find your problem and read about the options for corrective action. You, the contractor, can use it for quick reference on how to approach the job and owner, and you may also use it as a reminder on how to develop fair and equable contracts.

Good luck with the efforts you make to maintain the quality of the home regardless of its age.

Byron W. Maguire

1

ROOF AND CORNICE

OBJECTIVES

To restore the roof after it was damaged by a falling tree.
To restore the roof after wind damage to the shingles.
To restore the roof after damage by ice.
To renovate the roof after the life of the shingles or covering expires.
To understand the principles of roof construction.
To restore a damaged cornice after a severe storm or hurricane.
To understand wood cornice construction.
To understand aluminum and vinyl cornice installation.

OPENING COMMENTS

As contractors, we will be called by the owner to respond to an emergency when the roof or its cornice is damaged or worn out. We respond by examining the roof and its cornice, making onsite judgments, and discussing the renovations or restorations needed to make the assemblies like new or at least serviceable. Our goal is not to remodel the roof in any way, but merely to mirror the original architecture.

Items for discussion depend on the severity of damage or wear and tear on the roof and cornice. If the surface materials need replacement or restoration by adding new paint, we indicate that the job will be nominally difficult, although it could be somewhat expensive. On the other hand, if the damage has affected some structural members of the roof's framing and cornice, our discussion with the owner would be more in depth with the aim of alerting the owner to the necessity for extensive restoration and its accompanying costs.

In this chapter, we examine several renovation projects, beginning with the least difficult and progressing to the most involved. This organization will avoid repetition of the tasks associated with earlier projects. For example, we treat the weather-worn damaged shingle roof first. Since this project details the tasks related to removal of damaged shingles, flashing, and such, we know that more extensive restorations require the use of the same tasks; we do not need to repeat them. We will just enumerate them and refer back to the earlier project.

In every project we include a plan of action that includes setting out the specifications for the restoration in terms the owner can understand. Then we switch focus to the contractor and build a materials assessment followed by an activity list that provides estimates of time and a time line for construction. In the section on construction details, we add the principles of construction and also detail the tasks associated with each activity. Since there will be a wide variety of readers with varying degrees of skill and knowledge, some of the material will be extremely important and, at other times, you can skim over it or pick out only those parts that you need.

PROJECT 1. RESTORATION OF A ROOF AND WOOD CORNICE WEATHERED AND WORN OUT

Subcategories include wind-damaged sections of the roof; wind-torn or ice-damaged roofing, close and closed cornices, and gable end cornices.

Primary Discussion with the Owner

Problems facing the owner. As the contractor, called by the home owner, we must provide him or her with an understanding that ranges from little detail to a larger measure of detail from which we can provide a sound and reasonable estimate of work to be done and costs incurred. If we look at the condition of the shingles and their style, we can determine their condition using factors evident to the owner, such as shingles blown off the roof and shingles without any coating remaining. What may not be evident is the damage caused to the sheathing and cornice because of the water damage. We need to make a closer examination before the extent of damage and wear and tear is fully understood or ascertained.

We must use best judgment while discussing the damage with the owner. To understate the extent of damage to the owner gives the impression of an inexpensive solution. To overstate the damage or extent of restoration also sends an incorrect signal to the owner, who is now resistant to begin the restoration even though it is required.

The middle ground is the best. We must assess the extent of wear and damage and take some time doing it. Sometimes the owner or his or her representative can be shown the damages and thereby achieve a more informed understanding. This approach makes understanding the estimate of restoration and expenses easier to grasp and deal with. Psychologically, we are establishing the facts that the renovation has a cost that is related to the amount and extent of work and materials.

Sometimes different types of estimates work best. Sometimes an estimate written on the back of a business card works fine. Sometimes the estimate prepared on a proposal form with company name and logo works best. If the job is extensive, a snapshot estimate, even if totally accurate, creates a negative response, especially when details that must be included are omitted, whereas the proposal form with detailed items pertaining to the roof shingle's life expectancy and type, style, and kind, as well as a description of the replacement pieces for the cornice and painting systems to be applied, gives a clear answer to the owner.

Alternatives to the problem's solution. In this problem we must replace the entire roof even though the various slopes of the roof appear to have different degrees of wear. The south-facing slope has the most wear since the sun works its power for the majority of each day here. Shingles curl and slate covering erodes, and they are therefore subject to wind damage as well. Shingles on the east slope wear almost as quickly as the south slope, but not as badly. Shingles on the west slope also wear about the same as the east slope. But the shingles on the north slope show the least wear. However, a close examination of all the slopes reveals very brittle shingles throughout.

The owner would be mistaken if he or she decided to replace only the worst slopes when the life expectancy of the roof's covering has expired. In our best explanation, we must convey this requirement to replace the entire roof's shingles.

The cornice damage is different, and we can select one of several alternatives and present these to the owner. Cornices made from wood, as shown in Figure 1-1, consist of several pieces of lumber and plywood. They include the fascia, soffet, frieze board, and molding. Of these, the one piece most often in need of replacement due to weathering and water damage is the fascia. Its surface is weathered daily; the other pieces are sheltered, so they last longer.

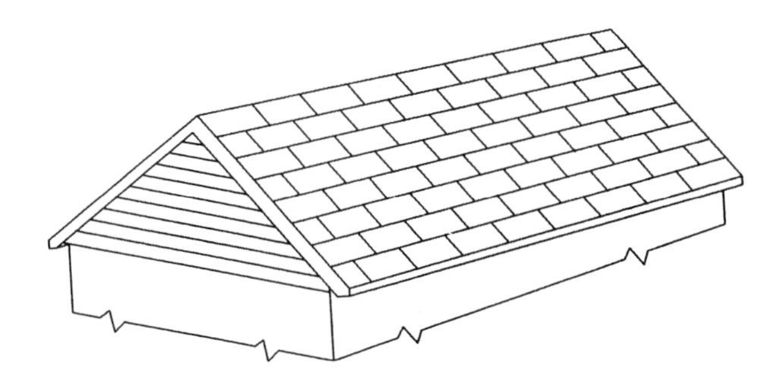

Figure 1-1 Shingle roof.

With today's painting systems, oil-based paints are not used nearly as frequently as the more modern latex types. Both are subject to damage, but from different causes, and present different problems, but require approximately the same corrective solution. The affected board, if decayed from housing water and growing fungus, must be replaced. Trapped water causes the oil-based paints to peel away. Trapped water in the board causes the latex paint to bubble, which may not necessarily rupture until probed with a screwdriver.

Stained boards weather differently and the wood cells actually erode. This causes boards to warp and split. The only cure is to replace the piece with like kind of lumber. But we are faced with the problem of matching old, weathered stain.

We must take into account all the above considerations before coming to a decision. The alternatives we provide to the owner depend on our close examination. Several are as follows:

1. Cornices with a painted system, but with damaged lumber and plywood, will have the damaged pieces removed and replaced with like kind. Then apply a new painting system to the entire cornice if older than 3 years.

2. Cornices with damaged painting systems, but no damage to the cornice lumber and plywood, will require paint removal by scraping or burning, sealing, and repainting. The extent depends on the age of the painting system, availability of a matching paint, and expertise of the painter.

3. Cornices with stained, but damaged, lumber will require that defective materials be removed and replaced with like kind. Then a matching stain will be applied. Where a match is deemed unlikely, the entire cornice around the house must be restained.

Statement of work and the planning effort. Our statement of work is clear: "*We are obligated to restore the home to its original character and quality.*" The color, style, and type of shingle must closely match the original shingle, especially as to life (that is, 20 years) and color and for strip shingles with a 5-in. exposed surface. With regard to the cornice our statement of work includes: "*the replacement of defective materials and application of the new painting system.*" Also an important item in the work effort is the need to include *"a 100% cleanup with no damage to gardens and shrubs, and no marks or damage to sidewalks and driveways."*

The planning effort is not all that difficult, but several usual items must be included. We must plan to estimate the job fairly and competitively in order to win the bid. We must integrate the job into the company's expected workloads and ongoing projects. Since we can supply an estimate, we know the extent of man-days needed to complete the work and the types of tradespeople required. We must plan for either a subcontractor for roofing and painting as well as some carpentry skills or have our own crews perform the work.

Contract. Since this work is relatively simple in terms of contracting, we can submit a proposal/contract combined document. The basic parts must be included. Since this is the first chapter of the book, we will define the elements of every legal and complete contract here. These are as follows:

1. *Parties and geographical location.* Names and locations of the principal parties and the location of the property where the work is to be done.

2. *Competent parties.* As contractors, we are one-half of the competent parties, and the building or homeowner or his or her legal agent is the other half of the competent parties who can enter into a contract.

3. *Lawful subject or object.* The fact that we are contracting to restore the roof and cornice to its original condition is a lawful subject. We must describe the subject or object in sufficient detail to isolate it to avoid misunderstanding.

4. *Legal consideration.* Both parties must receive legal consideration. As the contractor, we receive a check for payment. The owner in this case gets a renovated roof and cornice.

5. *Mutuality of agreement.* In this section of the contract, four parts must be considered and included as necessary:

- **a.** The offer of acceptance must be unqualified; there must be no terms or conditions (no coercion).
- **b.** With construction contracts, the bidding documents usually constitute agreement and form mutual agreement when accepted.
- **c.** Times, turnkey dates, payment schedules, and other schedules are elements of mutual agreements in construction contracts.
- **d.** Quality standards may also become elements of mutual agreements. These standards could pertain to materials and workmanship.

6. *Mutuality of obligation or genuine intent.* Both parties must truly and willingly intend to enter into the contract. Both parties must be on the lookout for invalidation. Honest mistakes can and should be worked out reasonably. Since we, the contractors, are submitting the bid/contract proposal, we must make clear the conditions that fall under this element. We must not fraudulently state intent, and we must expect the owner to have the money to pay for the work before the job starts.

For this job the contract looks like this:

Building Contractors
123 Main Street
State, ZIP
Ph 123/ 444 5555

Proposal for: Mr. and Mrs. Jones 1 August 199×
Address

Description and Estimate: Replacement of roof shingles. This includes the removal of the old roof and installation of new shingles with 20-year life span. Job includes complete removal of all waste and debris. The job also includes the replacement of fascia along the north wall of the building and repainting the entire cornice. This proposal is good for 30 days.

Total cost, tax included: $2300.00

Materials Assessment

Direct material	Uses/purposes
Strip shingles	Replace old shingles
15-lb Felt	Replaces old felt
Roofing nails	Hold felt and shingles
Roofing staples	Hold shingles to roof
Metal edging	Replace metal roof edging
1 × 8 Fascia	Replace the 48 ft of fascia
1 × 2 Trim	Replace the fascia trim
8d Galvanized nails	Nail fascia and trim
Painting system	Subcontractor—paint the cornice
Aluminum flashing	Flashing around the brick of the chimney

Indirect materials	Uses/purposes
Blocking behind fascia	Add support to rafter ends

Support materials	Uses/purposes
Tarpaulins	Cover ground and shrubs
Ladders	Access to roof
Carpenter tools	Construction
Automatic staplers	Nail shingles

Outside (contractor) support	Uses/purposes
Painter	Subcontractor

Activities Planning Chart

Activities	Time Line (days) 1	2	3	4	5	6	7
1. Contract preparation	×						
2. Scheduling and materials	×						
3. Cornice repair		×					
4. Removing roofing		×					
5. Installing new roofing		×	×				
6. Painting system		×		×			

Costs

Materials Costs	$______.__
Contractor/vendor costs	______.__
Rental and other fees	______.__
Total	$______.__

Reconstruction

Contractor support. In our first activity we have already discussed the contract preparation for this relatively straightforward job between us and the owner. We also need one between the painter and our company. For this contract our specifications to the painter would be as follows:

1. Prepare all the cornice for repainting by removal of old damaged paint.

2. Fill all defects or nail holes in cornice and seal with prime coat. Prime all new wood as well.

3. Apply two top coats of exterior latex semigloss enamel. Color to match the present color. Life span of paint system 10 years, rated.

The price will be added to our costs and is part of the total bid/proposal pricing.

Materials and scheduling. Before we can make sense out of the scheduling and materials requirements we need to have a better understanding of the work involved. Let's set some parameters.

The house in question has a simple A-framed roof design with a *6-in-12 pitch*, as the sketch in Figure 1-2 shows. There are the usual vents from the kitchen and bathroom, and we may need to replace the lead covers. There is also a brick chimney attached to one end, and it is flashed to the roof. We will need to replace the roof flashing as shingles are laid, but the flashing tuck pointed to the chimney seems good. Since the gable ends have louvers, we do not need to add louvers to the roof. We should add roof-edging metal up the slopes and across the north and south lower edges.

The roof with overhang is 32 ft long and the run of the house is 18 ft. We can calculate the dimensions of the slopes or use a framing square to aid us. Then we can determine the quantity of squares of shingles to order.

We calculate the quantity of square feet of roof as follows:

1. Determine the length of the sloping surface by multiplying the run (18 ft) by 6 in. per foot of run for a total rise of 9 ft. Then by using the *Pythagorean theorem,* we determine the length of the slope:

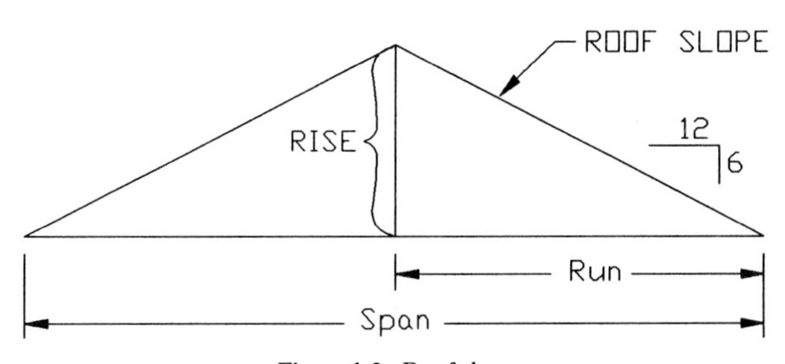

Figure 1-2 Roof slope.

Solution Using Figure 1-2, we see that the slope (s) is the hypotenuse, the run is a, and the rise is b.

$$s = \sqrt{b^2 + a^2}$$

$$= \sqrt{9^2 + 18^2}$$

$$= 20.125 \text{ ft or } 20 \text{ ft } 1\tfrac{1}{2} \text{ in.}$$

2. We can use the framing square *common-rafter table* and get the answer by using simple math. The scale under *6″ rise for L/ft run (in.)* is *13.42 in.* Since we know the run is 18 ft, the solution is as follows:

Solution: 13.42 in. × 18 (ft) = 241.56 in. or 20.13 ft.

3. Then multiply the roof length (32 ft) by the sloping length (20.13 ft) to obtain 644.16 sq. ft.

4. Then we double the quantity for both slopes and obtain 1288.32 sq. ft.

5. Then add starter strips at 11 per side for 22 pieces and ridge pieces at an average coverage of 15 in. per strip shingle for a total of 26 strip shingles [32 ft × 12 in./15 (inches of coverage per strip shingle)].

6. For allowance for waste, we add 10%.

Summary: Three bundles of shingles equal one square (100 sq. ft). There are approximately 76 shingles per square (or 25 to 26 shingles per bundle) in a 20-year rated shingle. Therefore, we need 15 squares of shingles.

Next we can simply determine the number of lineal feet of metal edging by adding the length of four slopes and two lower edges: 4 × 20.5 ft + 2 × 32 ft = 146, or fifteen 10-ft strips.

We also need 15-lb felt. One roll of felt covers 400 sq. ft, so we need four rolls.

We need roofing nails and staples for the staple guns and 1 gallon of asphalt cement.

We also need to provide 32 lineal ft of 1 × 8 fir for fascia and an equal length of 1 × 2 for trim for the north wall. These pieces must be nailed with 8d galvanized finish or casing nails.

Now we can determine the scheduling requirements. Due to the simplicity of this particular roof, an A-frame gable with a moderate pitch, we can strip and replace the roof in a short time frame. We set up the schedule with the owner. In this effort, we arrive at a date where the owner is ready to have the work done and has the funds for the work.

Then we examine the company schedule and look for the closest available date. Since we plan to use our own crew members for this job, we determine the number of roofers and carpenters needed for the work. In this job, one carpenter and roofer can perform the work by working together. We also can use a laborer to assist with the work, so we schedule one to drive the truck, pick up the materials, aid in stripping off the old roof, and provide general laborer work.

We must also schedule the painter. For this job, we call on a subcontractor and advise him or her of the job and estimate the time frame when the painting should be done. We make an offer for the work informally over the phone and receive acceptance. This "sub" customarily does our painting, sometimes on short notice.

Cornice replacement. The first activity under the at-the-site construction phase is the removal and replacement of the fascia on the cornice. Figure 1-3 illustrates the cornice from the end view and exploded view. Let's examine this detail to be sure we arrive at the same technique to replace the boards.

1. The gable end piece is nailed onto the fascia, and we must free the end by removing the nails. Since the nails commonly used are either case or common, we cannot drive them through. So we tap the board loose and then withdraw the nails.

2. Next we examine the fascia and notice that there is a 1 × 2 along the roof edge just below the shingles. This piece is nailed to the fascia. The fascia, itself, is nailed to the rafter ends. There is at least one joint in both the 1 × 2 and fascia. There is no metal roof edging.

3. We remove the 1 × 2 with a small crowbar or flat bar and hammer. No care must be taken to avoid breakage.

4. We remove the fascia by using a block of wood (1 × 2 by 6 in. long) and hammer. We tap the lower edge until the rafter ends are exposed; then we use the flat bar to pry the boards free. While doing this, we must be careful not to damage the frieze boards or plywood, since the piece's edge must fit into the new fascia.

At this point we either have bought pregrooved fascia or made our own at the job site. If we make our own, the groove must be 1/4 in. wide by 3/8 in. deep. Normally, this groove is placed 1/2 in. up from the bottom edge of the board. But, since we are matching the old board with a new one, we take our dimension from the old one. After cutting the groove with either a router with a 1/4-in.-square groove bit or by making

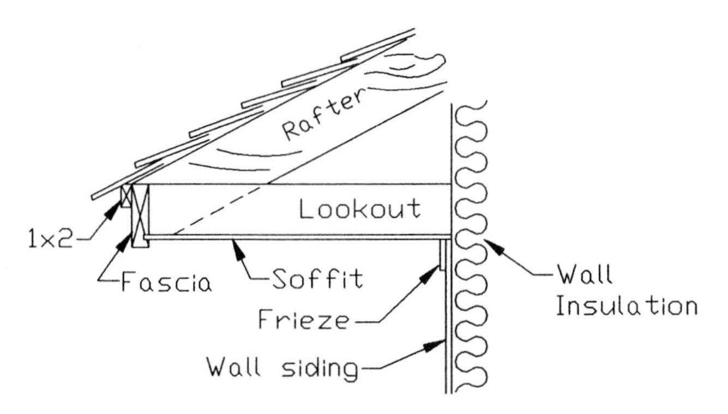

Figure 1-3 Cornice layout.

several passes with a portable power saw, we must trim the board's width to match the old board's width. This may be any dimension from 5 in. up to 7¼ in. Since the total length is 32 ft, we must prepare enough materials to cover the entire length and allow some for matching joints. The joints must overlap (45° miters) on rafter ends. So our best solution is to buy three 12-ft boards and make two joints.

We proceed with the replacement as follows:

1. Square the end of one board and place it next to the end of the fascia on the gable end. When in position and it fits, mark the other end for a miter, selecting the rafter closest to the boards end. Figure 1-4 shows the specific details that craftspeople use to ensure sound nailing. Notice that we hold the end of the miter back 1/4 in. from the rafter edge.

2. When the miter is cut, we place the board against the rafter ends and insert the soffet into the groove and nail it in place. Our nailing sequence should begin at the end that contacts the gable end fascia. The nails at the miter can be driven directly into the miter surface.

3. The second piece of fascia must have a closed and open miter. We begin with the closed miter and we place it over the open miter on the first piece. Then we make the end as before and make an open miter.

4. We install the board by aligning the miters for a perfect match and drive a nail back from the joint, yet into the rafter end. This is why we allow the 1/4-in. space shown in Figure 1-4.

5. The final piece in the 32-ft run must fit with a perfect miter at one end and a perfect butt cut at the other end. It is optional where we begin, so let's make the butt cut first by making a square cut at the appropriate end. Then we hold the piece in place and mark for the miter. After cutting the miter, we simply nail in place.

Construction note: Motion economy is very important. In the task just completed, we made several trips up the ladder and down, lifted boards several times each, and moved along the length of the cornice nailing the boards in place. If we set the nails as we drove them, we would have saved many retrace steps and practiced motion economy.

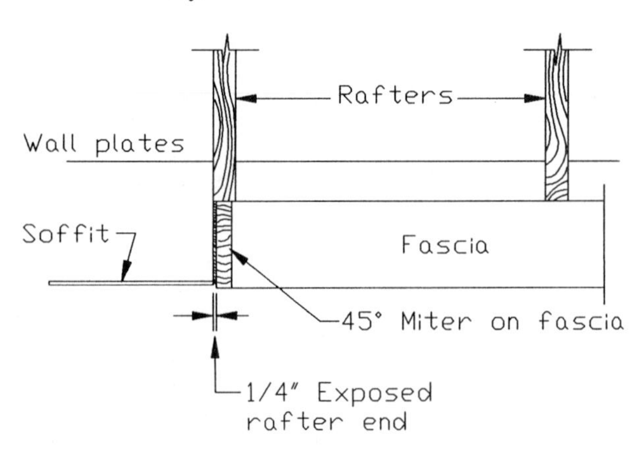

Figure 1-4 Cornice detail.

6. Replacing the 1 × 2 trim is simpler than replacing the fascia, but requires the same fundamental techniques. We cut and install each separately. However, we do this job in a shorter time since one of us cuts the pieces while the other nails them in place.

This concludes the cornice repair.

Removal of the roofing materials. The activity of removing the old roofing is not difficult in our generic project, since we can easily walk on the roof without the use of roof jacks. The tasks are as follows:

1. Remove the ridge pieces.

2. Remove the shingles.

3. Remove the flashing at the vents and along the chimney.

4. Remove the old felt (tar paper).

One of the best tools to use in removing the shingles is the shovel. A flat, square point or spade is the best selection. It is comfortable to work with since we can stand while using it.

First we pry up the ridge pieces. Although the job does not last long, when one of us operates from each slope we can pry up the entire length in a matter of minutes. This exposes the lap of shingles at the ridge.

Next we insert the shovels under the shingles at the ridge and pry them up along the ridge line. At first we pry up; later we either push down or pry up to free more shingles from the roof. Note that these shingles do not come off one at a time, but rather in bunches.

Since we must avoid damage to the landscape and sidewalks beneath the roof edges, the laborer must work to ensure that none of the loosened shingles and nails fall off the roof. Rather, he or she is tasked to pick them up and toss them onto the truck, which we have parked close to the building.

As we approach the lead flashing that covers the vents, we carefully remove the two nails holding each in place. At the chimney, we carefully pry each piece of flashing free from the roof with a flat bar, while ensuring no damage to the flashing tuck-pointed to the chimney. Figure 1-5 shows these pieces.

Once all shingles are removed from the slope, we remove the felt and examine the sheathing for damage or decay. Then we begin at the lower edge and install the 15-lb felt. We can install a single strip 32 ft long, but it is easier to manage in shorter lengths. Therefore, we will put two pieces along the lower edge with a 3- to 4-in. overlap. By using the lower edge as a reference, we have no trouble with alignment. We nail each piece at 12-in. intervals along the top edge, bottom edge, and once through the middle. Standard roof nails or special roof nails with large square heads may be used.

The second row of felt must overlap the lower row by 2 in., which is usually marked with white along the edge of the first piece. Again the overlap of the joint needs to be 3 to 4 in. The final piece of felt must overlap the ridge several inches. Then we repeat the process on the other slope.

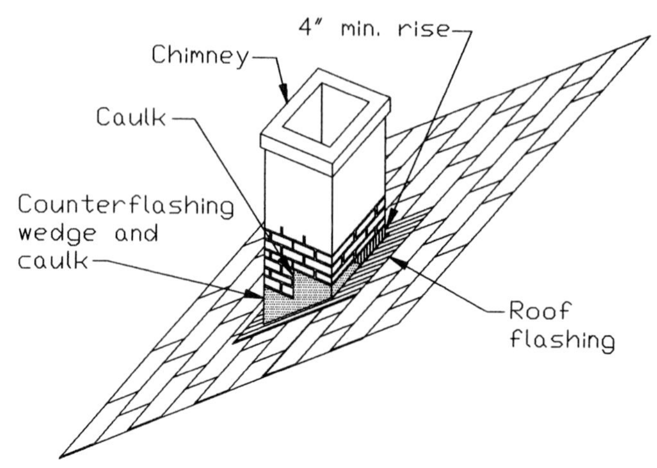

Figure 1-5 Chimney flashing.

Construction note: On a windy day we need to secure the felt better. Sometimes the best solution is to tack nail 1 × 2's along the horizontal overlap of the felt.

Installation of the new roof. The application of the new shingles is a straightforward series of steps. Most carpenters and all roofers know them well. Several rules must be applied.

1. A starter row must be installed along the lower edge of the slope.

2. The rows must be maintained at 5- to 5¼-in. exposure.

3. The vertical alignment must be straight and perpendicular to the lower roof edge, and alternate rows must be offset 6 in.

4. Shingles in the valley must meet or be an equal distance from the valley line.

5. Or the valley must be constructed with shingles extending across the valley to the other slope.

6. Flashing must be installed such that it aids in the fall of rain.

7. Ridge cap shingles must be equal distances each side of the ridge (6 in. each side).

In our generic project we are installing metal edging along the lower edge and slopes. These pieces are easily nailed onto the sheathing. They should be overlapped 2 to 3 in. Nails must be spaced about 12 in.

We must begin with the installation of the starter course by either buying rolls of starter or turning the standard shingle upside down. Since our roof will use the standard fiber-glass shingle, we can use for either method. However, if multilevel 25- and 30-year shingles are used, we should use the roll starter, since this allows for a much more even layer of shingle.

Next we prepare the shingles. We will cut 6 in. off a bundle of shingles and alternately nail a full shingle and cut-off shingle along the gable end of the roof. Each shingle will be nailed with three roofing nails or staples. Figure 1-6 shows the details. Then we can and should snap several chalklines horizontally as reference points to maintain alignment.

Since we are laying the shingles up the roof instead of across the roof, we will bring up two shingles at the same time. This economizes movement on the roof. One accepted practice these days is "four nails to a shingle," and another is to use "four and three nails." The four nails per shingle is simple to understand and requires that we lift up the full shingle when installing the next row to nail the endmost point. The four and three technique has us use four nails when we can without lifting any shingles and three when we must, as shown in Figure 1-7, to avoid lifting shingles. The second technique is a labor-saving technique since we can install more shingles per hour this way.

The second technique of strip shingle installation is with the stagger method. For this we start with a full shingle and decrease the length of each succeeding row

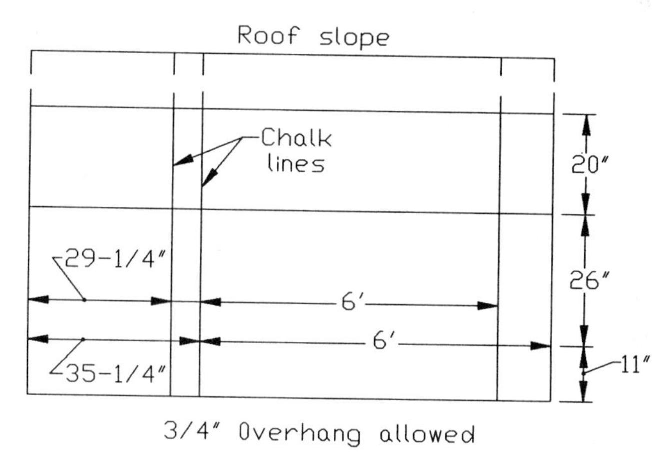

Figure 1-6 Layout for strip shingles.

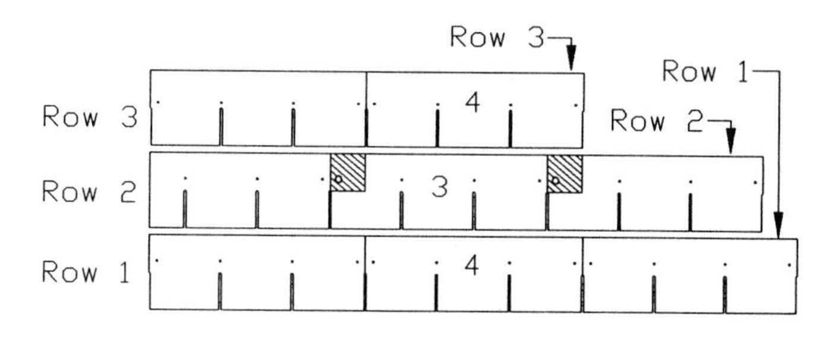

Figure 1-7 Applying strip shingles.

by 6 in. (36, 30, 24, 18, 12, 6), as Figure 1-8 shows. Then we start over again. When using this technique, we can install six rows in succession across the roof before beginning the next six rows. This technique requires considerable up and down movement across the roof. Thus the time for installation may be somewhat increased over the technique described previously.

Let's discuss the installation of shingles and flashing pieces. We will assume that during the removal process we were careful not to disturb the flashing tuck-pointed into the brick chimney. This means that we must cut several new pieces of flashing from a roll of aluminum or copper 12 by 12 in. Then, as shown in Figure 1-9, after installing the first piece past the lower edge of the chimney, we slide the flashing under the one on the chimney and press it flat onto the shingle. Then we lay the next row in place and nail it. Following this, we install the next piece of flashing. We repeat this process until the final piece is in place. After that, the final piece of flashing extends to the edge of the roof behind the chimney.

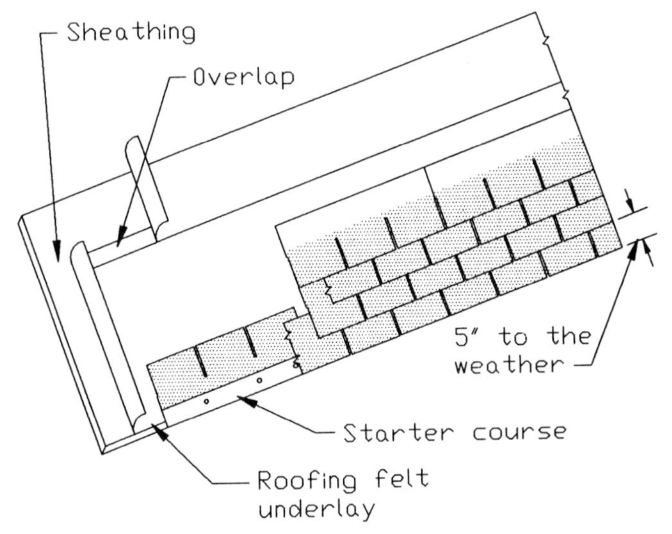

Figure 1-8 Strip shingle with 6-in. setback.

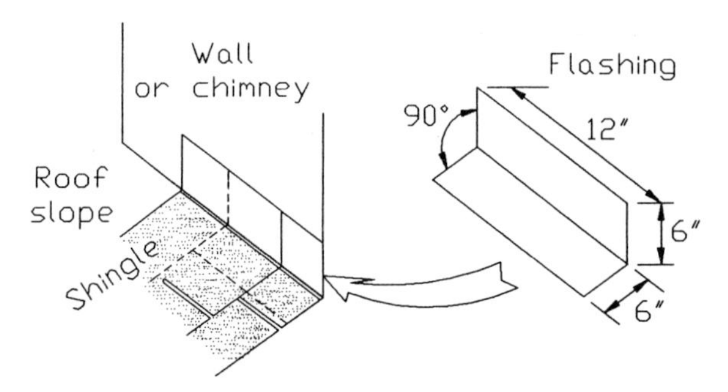

Figure 1-9 Roof/wall flashing.

The final row of shingles at the ridge line needs to extend over the ridge slightly, but never more than 5 in. The ridge cap only extends 6 in. either side of the ridge and must cover all overlap.

The ridge cap pieces are cut three 12-in. (pieces) per shingle. Each is nailed with an exposure of 5 to 5¼ in. The last piece is top nailed with four nails.

Painting. The final part of this job is the painting, which should be done in two steps. The first coat of paint on new wood should have been done when it was installed. However, it may not have been practical to do it then. Now, we contracted for removing all old and damaged paint from the cornices on all sides of the house. We can pressure wash some to remove old paint and mildew, scrape some where flaking has happened, or burn away the old paint with a torch. When the base is clean, free of decayed paint, and sanded, we must fill any holes, cracks, splits, or other openings in the cornice with a latex filler. When dry, we sand and apply a prime coat throughout.

Since we installed metal edging, we must prepare this metal for painting. It must be wiped clean to remove oil and cleaned further to neutralize the metal; then a prime suitable for metal is applied. The contractor's guide will detail the types to use.

Later we apply two top coats at 1.5 mils thick.

Concluding comments. This project illustrates the basics of replacement of a common roof with common strip shingles. We made the overall job more difficult by having fascia damage and old paint.

Anyone finding these conditions or similar ones should approach the job as we have done. This minimizes the efforts and costs.

PROJECT 2. RESTORATION OF A ROOF AFTER DAMAGE BY A FALLEN TREE

Subcategories include splicing and replacing the rafters; replacing the sheathing and shingles; restoring the wood cornice and painting; replacing damaged siding, damaged flashing, destroyed wood shingles, damaged wall plates, and maybe studs.

Primary Discussion with the Owner

Problems facing the owner. In this project, as sketched in Figure 1-10, we have been called to the house by the owner, who is in a state of anxiety over the tree in and on his roof (his could as easily be hers and is used here to denote either). The owner perceives extensive damage and faces the possibility of further damage if it should begin to rain again. What we see is the large limb penetrating the roof, crushed cornice along the gable end caused by part of the trunk, and lots of limbs resting on the roof. What the owner does not see is that the fly-rafter has been crushed and split, the cornice is destroyed along the gable end, a common rafter holding the fly-rafter lookouts is split, and maybe the second and third rafters are cracked or split. In

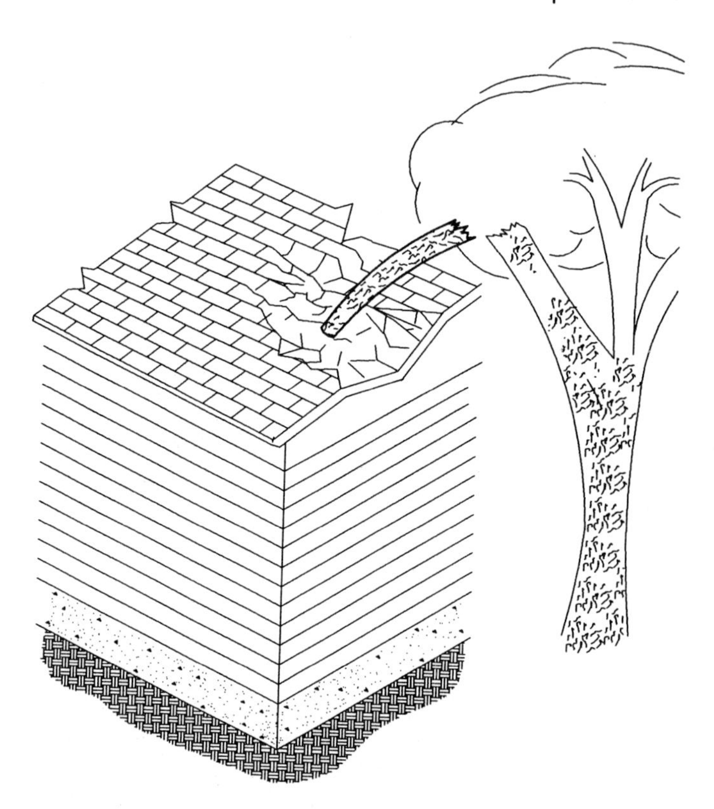

Figure 1-10 Damaged roof.

addition, he does not know that the sheathing is split and must be replaced. But, the owner does see that the shingles must be replaced. Upon examination from in the attic, we verify that three rafters are damaged; one is split in two, and two are slightly split. Upon further examination, we perceive that the conditions of the shingles shows some curling and the aggregate coating is very thin. This leads to the question of the age of the roof. We are told that it is 16 years old. What this means to us is that it will be impossible to match the shingles. Even if we could find the right color and type, the difference between old and worn shingles and new ones would prohibit an acceptable finished product. So we advise the owner to replace the entire roof covering.

Alternatives to the problem's solution. Our first effort should be to work with the insurance adjuster, since the owner has insurance. The adjuster will likely discuss the damage with the owner and make an offer. From our explanation of the extent of damage, the owner can speak from a position of knowledge when meeting with the agent.

With regard to alternatives, we can work within the constraints imposed by the adjuster to the degree that permits us to meet company quality standards. On the other hand, the adjuster may have provided us with the opportunity for more than one alternative.

In any event, with one alternative, we can replace damaged sheathing, cornice, and trim; splice rafters; and replace the entire roof's shingles. In a second alternative, we can replace the split rafters, since we already have made the decision to replace the entire roof covering.

We present these two alternatives to the owner with details and some range of costs. The owner elects to splice any rafter that is basically sound except for a split that does not extend through the entire width of the rafter.

Statement of work and the planning effort. This set of conditions permits us to develop a statement of work. *The roof will be entirely replaced*, since the age of the present roof does not permit our matching the shingles. We must *replace the fly-rafter and the first rafter inside the gable end* since the tree limb cracked both through. The third rafter is only slightly split, so, we can *splice the cracked rafter.* We know that *the cornice, which is a closed design with 12-in. overhang, must be replaced.* This means that the fascia, soffet, and frieze boards must be replaced and painted. In addition, we must *replace the sheathing around the area where the tree contacted and penetrated the roof.* We also contract to *clean up the site* by removing the tree.

The planning effort began with the collection of specifics about the rafter dimensions, cornice construction, and type of specifications to include in the bid proposal. One of the first contacts we need to make is with a tree-removal company to remove the tree. We will be there to seal the roof to prevent further damage. Further planning includes determining that two carpenters are needed to make the restoration. A two- to three-person roofing crew is needed for the roof's replacement. A painter will need to match the house paint. Due to the urgency of the situation, we will schedule the job as soon as possible after the contract is signed.

Contract

Building Contractors
123 Main Street
State, ZIP
Ph 123/ 444 5555

1 August 199×

Proposal for: Mr. and Mrs. Storm
Address

Description and Estimate: Restore the roof and cornice to its original design and condition. Replacement of rafters cracked through and splicing others split but only partway. Replacement of damaged cornice and damaged sheathing. Replacement of roof shingles with 20-year strip shingles. Tree removal and property cleanup associated with the damage and repairs. Painting new materials and entire cornice on the one gable end to match the house colors.

(*continued*)

Estimated costs: $5500.00

This bid remains effective for 30 days and includes the emergency repairs and tree-removal service costs. If the contract is not entered into within 30 days, the cost of tree service and emergency repairs, $350.00, will be paid in full by the owner at that time.

Materials Assessment

Direct materials	Uses/purposes
20-Year shingles	Replace worn/damaged roof
15-lb Felt	Underlayment for roof
1/2-in. Sheathing plywood	Replace the damaged sheathing
1 × 8 fascia	Replace the broken ones
1/4-in. Plywood	Replace soffet
Molding	Replace frieze
2 × 4's	Replace fly-rafter and lookouts in fly-rafter assembly
2 × 8	Replace rafter and splice rafter
1/2 × 4½ in. Bolts	Hold splice in place
16d, 8d, common and 6d finishing nails	Hold members in place
1 × 2 Cornice trim	Trim along the roof edge
Roofing nails or staples	Nail shingles and felt
Caulking	Seal joints
Painting system	Exterior latex painting system for cornice

Indirect materials	Uses/purposes
4- or 6-mil Plastic	Emergency repairs to seal the roof
1 × 2	Emergency repairs to hold plastic
Assorted nails	Emergency repairs

Support materials	Uses/purposes
Metal scaffolding	Reconstruction efforts
Roof jacks and planks	Aid in roof work
Carpentry tools	Construction
Roofing tools	Reroofing
Ladders	Transporting materials and workers
Table saw or radial arm saw	Eases work
12-in. House jack	Straighten out sagging rafter

Outside (contractor) support	Uses/purposes
Tree-removal service	Remove the tree from the house and property
Painting subcontractor	Painting support

Activities Planning Chart

Activities	Time line (days) 1	2	3	4	5	6	7
1. Emergency repairs	×						
2. Remove damaged materials		×					
3. Replace rafters and sheathing		×					
4. Rebuild cornice			×				
5. Prime and paint cornice			×	×			
6. Replace the roof covering			×	×			

Reconstruction

Emergency repairs. The fundamental purpose for the emergency repairs is to prevent further damage to the roof and interior. Added rain would surely spoil more of the ceiling and insulation above the ceiling.

We assume that the tree-removal service has completed the work of cutting the tree and removing it from the house and roof. This allows us to more closely examine the damage and effect temporary repairs.

The cornice has been crushed by the tree trunk, but surprisingly the gable end wall is intact. A major limb broke off while contacting the roof and its stub penetrated the sheathing and split two rafters. Another limb struck the roof and caused the third rafter to split.

To make the emergency repairs we must *dry-in* the roof and cornice. We sweep off the roof to get rid of the leaves and bark. Then we remove the shingles around the hole made by the tree and further damaged by removing the tree limb.

The best way to dry-in the roof is to patch the opening with 1/2-in.-thick sheathing plywood, which we do. Then we cover the plywood with several layers of vinyl sheathing. We make sure to install the vinyl under a layer of shingles above the damaged areas to ensure that any rain will run over the damaged area. To ensure a sound seal, we nail 1 × 2's along the sides and across the bottom of the vinyl.

Since the damage extends to the outer edge of the overhang, our patch job extends there, too. By extending the cover this far, we are able to protect the gable end from rain leakage that could blow in behind the siding.

For this project we use roof jacks since the pitch is *8-in-12* and it is not safe to walk on. These jacks should be installed about 2 to 3 ft below the damaged area to make working easier and safer. We must use the ladder to get onto the roof and alongside the gable end to inspect for damage to the siding.

Removal of damaged materials. The variety of tasks we must complete are as follows:

1. Remove the roof shingles and felt from eave to ridge and from gable end edge to the rafter past the third one, which we determined is cracked.

2. Cut away the sheathing from eave to ridge, over half the thickness of the split rafter since we are only splicing this one.

3. Remove the sheathing. This exposes the rafters.

4. Remove the fascia, frieze, and soffet from the overhang.

5. Cut the rafters to be replaced in several pieces for easy removal.

The reason why we remove only sufficient shingles to expose the rafters we need to work on is to minimize the possibility of further damage should it rain. By stripping the shingles off the roof one rafter's spacing past the split rafter, we make access simpler and safer to work around.

The reason why we need to cut the sheathing over the split rafter, leaving the sheathing half-lapping the rafter's thickness, is two-fold. First, there is no need to remove undamaged sheathing. Two, we can easily splice the rafter with only one-half exposed. This technique also provides the 3/4-in. nailing surface we need to nail the replacement sheathing, too.

We could try to remove the rafters next, but because of the fly-rafter assembly construction, as shown in Figure 1-11, it would be inconvenient. Therefore, we first strip away the gable end cornice from lower eave to ridge. This exposes the fly-rafter and, with the sheathing removed, all the rafters we need to repair and replace are readily seen.

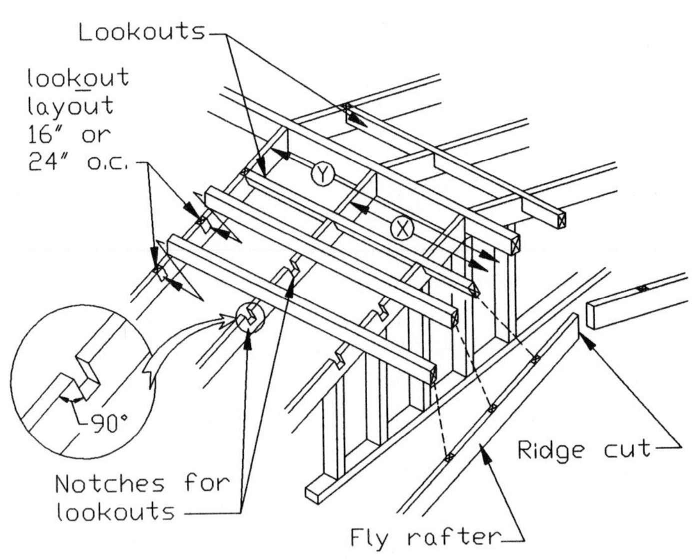

Figure 1-11 Fly-rafter design.

Finally, we remove the fly-rafter and the two broken rafters. While removing the rafters, we must remove the 2 × 4 lookouts as well. If it is possible to save some of the lookouts, we can do so, but since they are only a 2 × 4, there is little cost, and it would be simpler to not reuse them. Once the lookouts are removed, we can remove the broken rafters by making several cuts through first one and then the other. By removing the center section of rafter, we can more easily free the section from the ridge board by pressing the rafter down and twisting, while using the prybar to pull it free.

We can free the lower end that has the bird's-eye cut and fits into the lower eave cornice, but this job is more difficult and requires care or the lower eave cornice will be damaged. Here is one way we can proceed. We can cut away the rafter at the bird's mouth. Then we can drive the nails, in the fascia, through and into the rafter end by using a nail set. Then we can insert a flat bar between the rafter overhang and the lookout nailed to it and gently pry it loose. Normally, the lookout is nailed to the wall and the soffet is nailed to it. Therefore, it should remain in place.

Replacing the rafters and sheathing. The work involved in this phase of the project is complex, and requires a knowledge of rafter layout and installation and the construction of the fly-rafter assembly. The installation of the sheathing is a simple task, but one that requires safe practices.

For this project, we must work from the cracked rafter outward to the fly-rafter assembly. Therefore, our first effort is to splice the cracked rafter with an 8-ft piece of rafter material, a 2 × 8. The tasks are as follows:

1. Realign the rafter to take the sag out of it.
2. Prepare the splicing piece by cutting it for length and positioning it.
3. Securing the splice.

The simplest way to take the sag out of the split rafter is to use a 12-in. house jack. We lay a piece of 2 × 8 flat across at least three ceiling joists, plumb or under the center of the split area in the rafter. This only has to be approximate, not exact. Then we set the jack on the 2 × 8 and cut a short 2 × 4 to fit from the top of the jack to the under side of the rafter. Then we tack nail the 2 × 4 in place with a 12d common nail.

Next, as Figure 1-12 shows, we must establish a reference so that we know when the rafter is realigned. We use the mason line for this effort. At the ridge, we nail a scrap 1 × 2 to the top edge of the rafter at the ridge and let the nailhead stick out about 1/2 in. We tie one end of the mason line to the nail making sure that the line touches the 1 × 2. Then we go to the lower eave and repeat the process. But, we must pull the line tight and tie it without slack. This line establishes our reference.

Now we take the electric drill with 3/8-in. bit, an 8-ft 2 × 8, bolts, washers and nuts, and wrenches up on the roof. While one of us raises the house jack from the ceiling area, the other places another piece of 1 × 2 between the rafter's top edge and mason line. When the line touches the 1 × 2, we are exactly in alignment. However, all rafters sag because of the weight of the roofing and sheathing; therefore, we slide the loose piece of 1 × 2 back from under the line and raise the rafter another 1/4 in.

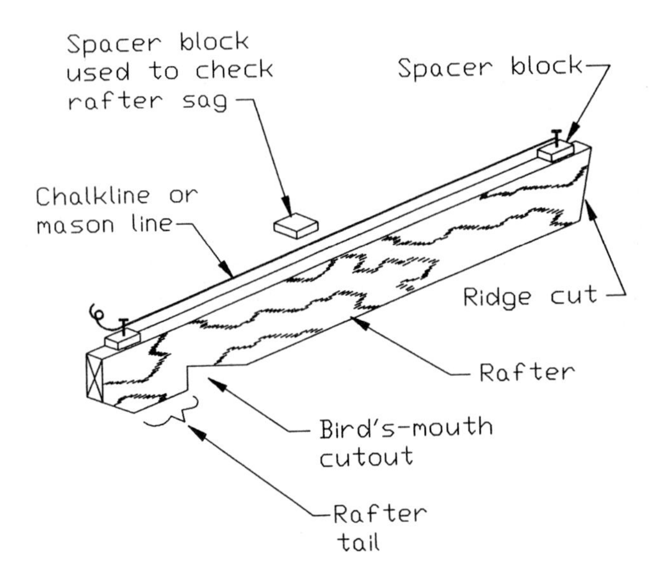

Figure 1-12 Rafter with special cuts.

With the rafter in position, we now install the 8-ft splice. While positioning the piece, we drive several nails to hold it up. Then we drill eight holes about 2 ft apart, in pairs, through the splice and rafter. Into each hole we install a bolt and secure it with a wrench. There must be a washer between wood and bolt head and wood and nut. This completes the splicing operation. We remove the jack, short 2 × 4, and mason line with blocks of wood.

Our next task is to replace the two broken rafters with new ones. This is not a difficult task for qualified craftsmen, but deserves some explanation for those not informed. We know that the rafters are cut from 2 × 8 number 2 common or machine-stress rated fir or yellow pine. We also know that every 2 × 8 has a crown edge and that the crown must be positioned toward the sky or up. Thus we position the rafters on the workbench and mark the crown edge with a pencil ×.

Since we have already determined that the pitch is 8-in-12, we lay out the ridge cut with the framing square as shown in Figure 1-13. Or we can use the end of the old rafter as a pattern. Either technique produces the same results, which is a line for trimming the end of the rafter. The longest point is the top or crown edge.

Next, we measure the length of the rafter by taking the measurements from the roof itself. Therefore, one of us holds one end of the tape measure at the ridge board and the other end at the top-back edge of the fascia. This length must be transferred to the 2 × 8 on the workbench. This time the mark is the short point, and the square is used once again to make the 8-in-12 line, or we can use the old piece to mark the cut. While we are at it, let's make the bird's-mouth cutout as well. For this, we definitely use the old rafter as a pattern.

All the lines are made; now we use the portable power saw to make the cuts. While making the bird's-mouth cut, we will finish the cutout with a handsaw. Since we need two new rafters, we mark the second one with the first one and cut it.

Installing these rafters is a straightforward task. One of us nails the rafter to the ridge board and the other nails it to the wall plates at the bird's-mouth cutout. Then,

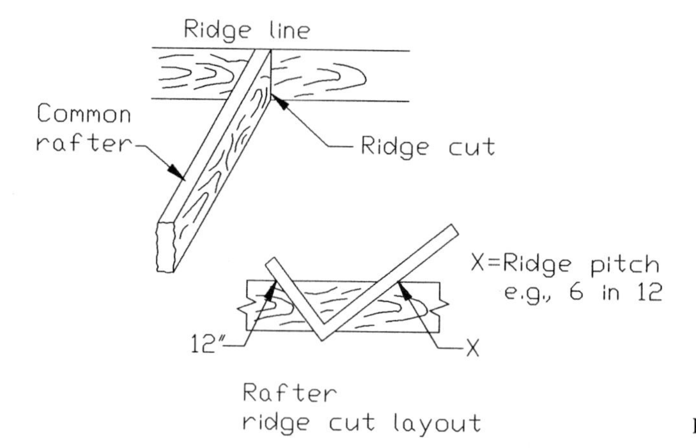

Figure 1-13 Rafter ridge cut.

we also nail it to the lookout and drive two 8d finish nails through the fascia into the rafter end. This completes this phase of the restoration.

Reconstructing the fly-rafter assembly. The old assembly that was destroyed by the tree was made from 2 × 4 materials. We will restore the assembly by using the same type of materials. The assembly consists of a single 2 × 4 fly-rafter and short (approximately 3-ft-long lookouts spaced about 24 in. o.c. (o.c. means *o*n *c*enter). The fly-rafter has the same end cuts as the common roof rafters. So we can use the same length as for a common rafter.

Since we have the rafter on top of the wall's gable end intact, we will use the cutouts in it as references to mark the points on the first common stud where the lookouts must be nailed. For this task, we use the framing square and make the marks. Then we measure the length of the old lookout and cut sufficient new ones to replace each old one. Next, we nail these in place with 16d common nails through the common rafter into the ends of the lookouts and 8d common nails using the toe-nailing technique into the rafter with the slots cut out. Before using the 8d nails though, we must install a mason line along the outside tip from the ridge to fascia where the inside edge of the fly-rafter must be. With this line in place, we will be assured of an exact straight line along the ends of the lookouts. First, we position the lookout closest to the middle of the length of the fly-rafter length and nail the lookout. Then we work both up to the ridge and down to the fascia.

We next nail the fly-rafter in place with its top edge flush with the lookouts. This completes the fly-rafter assembly.

Next, we install sheathing on the roof. For this we use CD-rated sheathing with a thickness to match the old roof's sheathing. If the roof were sheathed with 1 × 6, we would use 1 × 6. If the roof were covered with wood shingles, we would replace the 5/4- or 1 × 3 nailers to match the old ones. The sheathing must be trimmed flush with the outer edge of the fly-rafter. Nailing along the edges must conform to standards at 6

CD refers to the quality of outer layers of plywood. C quality has some unfilled knot holes; D quality has many unfilled knot holes.

in. o.c. The nailing throughout the centers of the sheathing must by 8 in. o.c. Between each sheet of plywood we install H-clips to prevent sagging between rafters.

At this stage of the renovation, we must dry-in the roof with new felt.

Cornice reconstruction. In keeping with our goal to restore the home, we pay close attention to the construction of the old cornice on the gable end. We have three pieces to replace, the fascia, the soffet, and the frieze board or trim.

To prepare the fascia, we must use a short scrap piece from the old, broken one. We need two things from this piece: (1) the width of the piece, and (2) the position and dimensions of the grove for the soffet. Our first task is to trim the piece for the total width, and we do this with a portable power saw or bench saw if one is on the site. Then we mark the position of the groove and set the saw's guide for the first pass through the board. We set the blade for the 5/16-in.-deep cut and make successive passes until the groove is slightly more than 1/4 in. wide. (This permits easy installation of the 1/4-in.-thick soffet pieces).

The actual installation only requires that we position the piece flush with the top of the sheathing before nailing it into the lookouts and ridge piece. A quick observation of the position of the groove will indicate that it is placed exactly flush with the bottom of each lookout. Before we can do that, we must cut the bevels at the ridge end and lower end. The one at the ridge must be cut on the ground. However, we can easily cut the waste from the lower end after the piece has been nailed in place. Most contractors do this, thereby ensuring a perfect fit. On some cornices, we add a flare piece to account for the cornice at the lower end. Figure 1-14 shows us what this piece can look like.

We plan to use 1/4-in. plywood as soffet, just as was originally used. We can use an old piece as a template for width and rip the sheet into several strips. We will need at least two, and the splice between them will be a butt joint that is joined at a lookout. The installation is simple, but the working area is unsafe. We will be working with ladders. We slide the first piece into the groove and slide it up to the ridge. Then we nail it to the lookouts with 4d finish nails. Then we fit the second piece as required and nail it into place. This completes two-thirds of the work.

The frieze board or trim are the last to be installed. Again we must work from ladders. The board or trim should be installed from the ridge down. Butt joints are customarily used, but a miter at the splice end makes a more professional job. The frieze board or trim is nailed to the siding with 8d finish nails.

This finishes the renovation of the structure and trim.

Priming and painting the cornice. As soon as practical, the painter must apply a prime coat to the raw wood. This task only takes an hour or two at most. So it is a task that employs less than a full day. The subcontractor will be reluctant to send a worker; however, sending one person near the end of the workday is reasonable. The painter should apply a prime coat of latex or oil base depending on the current painting system used on the rest of the house. This information is readily available from the owner, who has probably repainted the cornice more than once. This undercoat can be white.

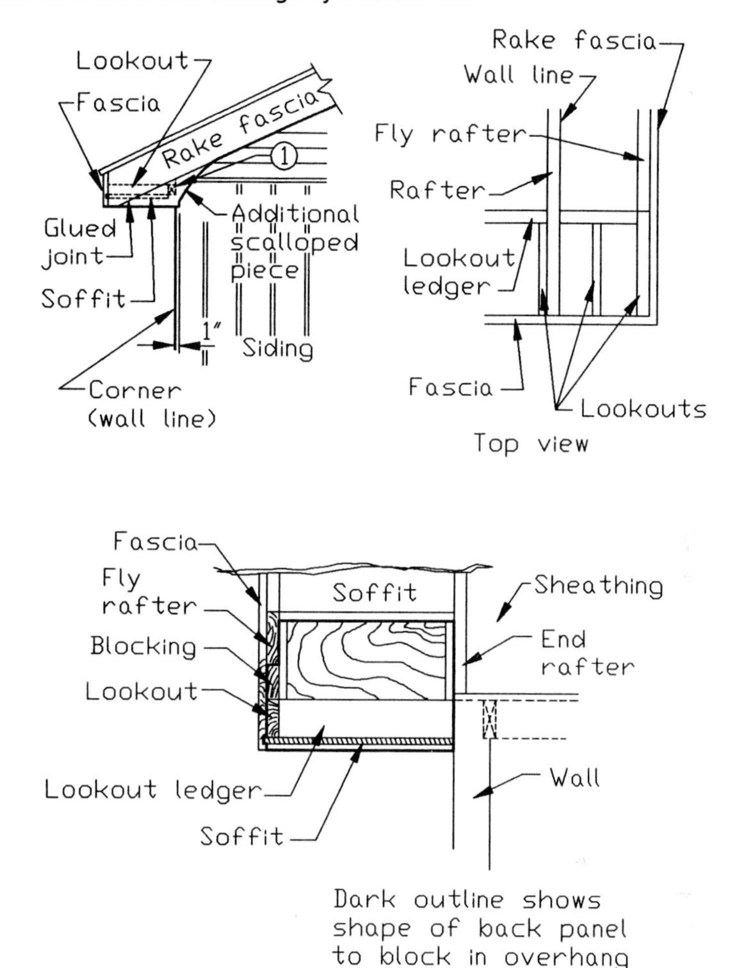

Figure 1-14 Gable-end overhang and cornice.

Since the subcontractor now knows that the work is ready for painting, he or she needs to send the painter back to the house to apply the remaining top coats to finish the job. In the activity chart, we show the entire job taking 2 days. The prime coat was done on the first day, and the other two coats were done on the next available day, with one coat first thing in the morning and the last coat at the end of the day.

In this situation, we assumed that the house paint was new enough for the painter to match the paint for type and color.

Roofing. Since this roof is a gable end style, we do not need to cover the discussion of reroofing. There is nothing unusual or different from what we examined in project 1.

Concluding comments. In this second project we learned how to approach the restoration of a roof that was damaged by a fallen tree. We protected the house from further damage while negotiating the contract. We replaced damaged and broken members, spliced a cracked rafter, rebuilt the fly-rafter assembly, and restored the cornice and roof to their original condition.

PROJECT 3. RESTORATION OF HIP ROOF, CLOSE, AND ALUMINUM CORNICE, PHYSICAL DAMAGE TO THE FRAMING CAUSED BY SEVERE WEATHER OR A HURRICANE

Subcategories include removing the hip rafter framing and replacing with new members; rebuilding the cornice.

Primary Discussion with the Owner

Problems facing the owner. In this problem the contractor recognizes that the hip roof has been severely damaged by a storm. In the corner that received the most damage, hip jacks and the hip were broken. The contractor also recognizes that the cornice is a *close* type, which is made from wood and covered and molded with bonded aluminum, as shown in Figure 1-15. The discussion with the owner brings out the following problems. The owner is told that the hip rafter supports the hip-jack rafters nailed to it, and the hip and some of the jacks are broken. The only possible

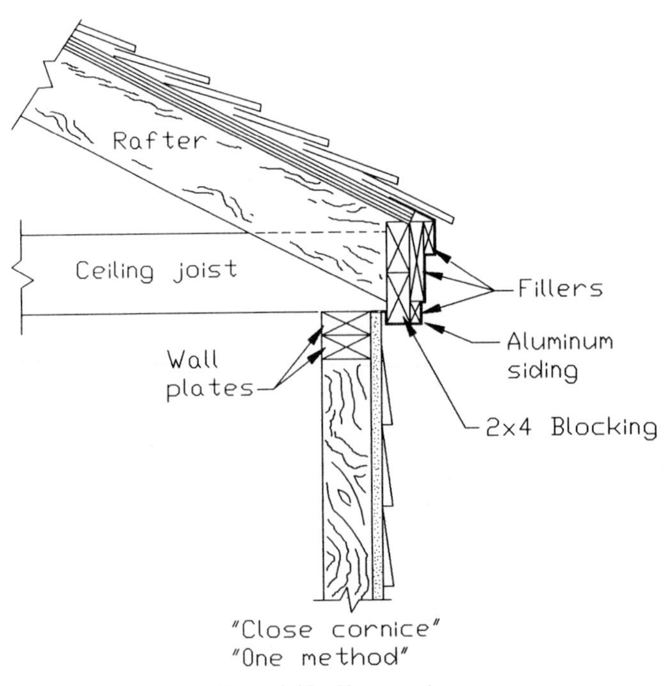

Figure 1-15 Close cornice.

solution is to replace them. Since the rafters are broken, the sheathing and roofing are damaged as well. When the roof was struck, the wind also played havoc with the cornice, due mostly to the raising action from high winds. Since the cornice was built of wood and covered with aluminum, the aluminum bent. We also discuss the fact that the repairs must be made to the two-story house and that we will need to install scaffolding on two sides of the house.

Alternatives to the problem's solution. There are not many alternatives to this restoration project. We must replace all broken framing members, roof sheathing, roofing, and cornice. Since the aluminum is bonded with paint, we will not need a subcontractor for painting, but we will need an aluminum and vinyl siding contractor. The roofing is only several years old, so we can match the type and color of shingle. But the several years have changed the color some, and there is a little erosion on the shingle surface exposed to the weather. We could just replace the damaged shingles, but the owner prefers to have us replace the slopes that were damaged. We agree with the decision.

We could work from ladders with a scaffold board between them, but for safety reasons and because of the amount of materials we will need to remove and replace, we option for rental scaffolds.

Statement of work and the planning effort. We must *remove the damaged members from the roof*, which include the hip and jack rafters on two slopes. We must also *replace the sheathing broken in the storm*. Since the cornice was damaged, we must *replace the cornice on two sides of the house*. We agreed to *replace the shingles on both slopes* damaged by the storm.

Contract

Building Contractors
123 Main Street
State, ZIP
Ph 123/ 444 5555

1 August 199×

Proposal for: Mr. and Mrs. Windstorm
Address

Description and Estimate: Restore the roof and cornice to its original design and condition. Replace hip and jack rafters cracked or broken through and damaged sheathing. Replace damaged cornice and aluminum cornice covering. Replace roof shingles with 20-year strip shingles on both slopes that sustained damage. Property cleanup associated with the damage and repairs is included.

(*continued*)

Estimated costs: $10,750.00

Method of payment: 25% ($2600) on contract signing; balance due on completion of contract. *Exception*: Upon notice that the repairs will be covered by homeowner's insurance and pledged to us, no deposit is required.

This bid remains effective for 30 days and includes the emergency repairs. If the contract is not entered into within 30 days, the cost of emergency repairs, $750.00, will be paid in full by the owner at that time.

Materials Assessment

Direct materials	Uses/purposes
2 × 12	Hip rafter
2 × 8s	Hip jacks
Sheathing	Cover the rafters
15-lb Felt	Dry-in the roof
25-Year shingles	Roof covering
Cornice wood	Lumber, trim, and molding
Aluminum	Cornice
Assorted nails and staples	Secure lumber and shingles
H-clips	Used with sheathing

Indirect materials	Uses/purposes
Insulation	Replace damaged material
2 × 4s	Temporary bracing
6-mil Vinyl sheet material	Dry-in the roof; emergency repair
1 × 2 Strips	Hold down vinyl

Support materials	Uses/purposes
Rental scaffolding	Establish working area above the ground
Scaffold planks	Walking surface for scaffolds
Carpentry tools	Construction
Power stapler	Install shingles
Power saws	Cut lumber
Workbench	Work surface for preparing rafters
Pulley and rope (optional)	To raise materials to roof
Electrical drop cord	Power machines

Outside (contractor) support	Uses/purposes
Siding contractor	Install new aluminum siding

Activities Planning Chart

Activities	Time line (days) 1	2	3	4	5	6	7
1. Emergency repairs	X						
2. Install scaffolds		X					
3. Remove damaged materials		X					
4. Replace framing		X	X				
5. Sheath and dry-in roof			X	X			
6. Replace cornice wood				X			
7. Replace cornice aluminum					X		
8. Replace shingles					X	X	
9. Job site cleanup						X	

Reconstruction

From the activity chart, we observe the usual arrangement scheme for this project. In addition, we must state that the above plan is designed for two craftspeople on site throughout the project. If we increase the work crew by one more person, we can decrease the time frame by 1 to 2 days. The subcontractor will probably bring help since the cornice covering is molded on the site and requires more than one to hold it in place and fasten it.

Emergency repairs. The emergency repairs are performed parallel to those in the earlier projects. We must close the holes in the roof to prevent further damage should it rain again. For this effort, we use the extension ladders to reach the roof and vinyl to close the holes in the roof. The 1 × 2s should hold the plastic in place. We nail the aluminum cornice covering back in place to prevent damage to other parts of the siding and windows.

Install scaffolds. Rental scaffolds are welded assemblies with braces that clamp the end sections together. They telescope onto each other as we build upward. Each end section has at least two horizontal bars that we set the planks on to form the walking surface. For this project, we need some sections along the side where the roof is damaged and along the front where there is also damage. Just how far across the wall we install the scaffolds depends on the on-site conditions, pitch of the roof, and climate in which we must work.

The planking used must be at least two boards wide. This makes a 22-in.-wide work and walk surface. We always nail a scab of 1 × 6 across the planks about midway through the span. This prevents the boards from bending independently when we walk on the surface and thus adds a great deal of safety. In addition, a safety bar must be bolted onto the back of the last or top sections to act as a safety factor.

Metal scaffolds erected properly do not need to be secured to the wall, but many carpenters tie a line to the house as an added safety measure. With the scaffolds in place, we can proceed to remove the damaged materials from the roof and cornice.

Removing the damaged materials. As we did earlier, we only expose the roof where necessary to minimize further damage. However, this project is much more difficult since the hip rafter was damaged and it must be removed and replaced. The tasks are as follows:

1. Remove the old shingles and felt.
2. Remove the plywood sheathing.
3. Remove the cornice through the area of exposed roof on both slopes.
4. Remove the jack rafters (see Figure 1-17).
5. Remove the hip rafter.

Removing the shingles was covered earlier, as was removing the sheathing.

Exposing the cornice was also covered earlier, but it was the *closed* type with a 12-in. overhang. This time the cornice is a *close* type and that makes a difference. Look again at Figure 1-15 and find the following details: (1) 2 × 4 wall plates, (2) 2 × 4's nailed to the end of the rafters, (3) the position of the end of the rafter to the wall plates, (4) the relationship of wall covering and siding to the cornice, and (5) the wood that gives shape to the cornice.

To remove the cornice, for this project, we should cut the aluminum at a point beyond the common rafter on each wall. Then we should remove the pieces that give shape to the cornice. This exposes the 2 × 4's. We must remove the 2 × 4's as well, but we may need to cut them to do this. If the natural joint is not at the common rafter, draw a line along the face of the 2 × 4's such that it bisects the common rafter end (3/4-in. overlap). After pulling out the nails from the 2 × 4's, we use a portable power saw to make the cut. This, then, removes the last obstacle; now we can remove the 2 × 4's.

Removing the rafters begins at the corner. We start with the shortest jack rafters first and progress into the roof. If any rafters can be saved by careful practices, we can use them again. However, any split or damaged rafters must, by contract, be replaced. The next two rafters to remove are the common ones. Finally, we can remove the hip rafter.

Note: For those who are not familiar with the terms *jack rafter* and *common rafter*, let's define each. A jack rafter is one with a common rafter cut on one of its ends and a hip jack cut on the other end. This means that one of its ends fastens to the wall plate (hip jack) or roof ridge board (valley jack). The common rafter is one that extends from wall plate to ridge.

This completes the removal phase. All materials removed from the roof and cornice must have the nails taken out, even if the materials will not be used again. Safety first.

Replacing the framing. We lay out and cut the rafters on the ground, pass them up to the roof, and install them. Cutting the hip and jack rafters is not difficult when we use the portable saw, since we set the saw for a 45° miter and cut along the line on the rafter end. But laying them out and determining the length of each is a complex task that only experienced craftspeople can perform.

In our renovation situation we have the old, but damaged, rafters to use as patterns. We lay the jack rafter alongside the new rafter materials and mark both the plate cut and hip-jack cutting line. Then we cut the hip-jack cut along the line with the saw at 45°, but the plate cut is a straight cut. One cut fits on the plate and the other fits plumb with the wall and is ready for the 2 × 4 base of the cornice. Since the damaged roof in our project is an equal pitched roof, we must make the hip jacks in pairs. This means that one fits onto one wall plate and its partner fits onto the adjacent wall. When they are nailed to the new hip, they will match as Figure 1-16 indicates. Figure 1-17 shows one pair of jacks.

Since the hip was broken by the storm, we can use it for layout of the end cuts. However, we must verify the total length with a measurement from the roof. This verifies the length. In making the cuts to fit against the common rafters at the ridge, we make two 45° cuts as Figure 1-18 shows. The plate cut has the flat side of the bird's mouth and two vertical cuts at 45°. These cuts differ from the ones on the jacks because the jacks are 2 × 8 material and the hip is a 2 × 12.

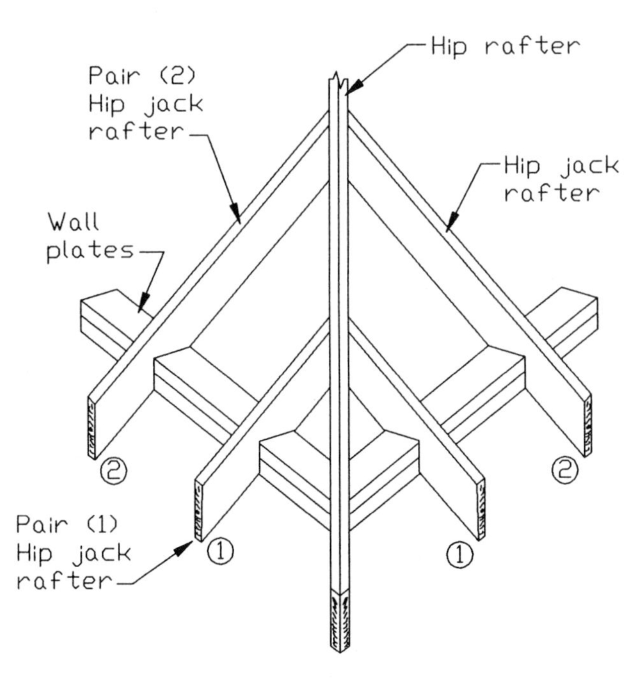

Figure 1-16 Hip and jack-rafter assembly.

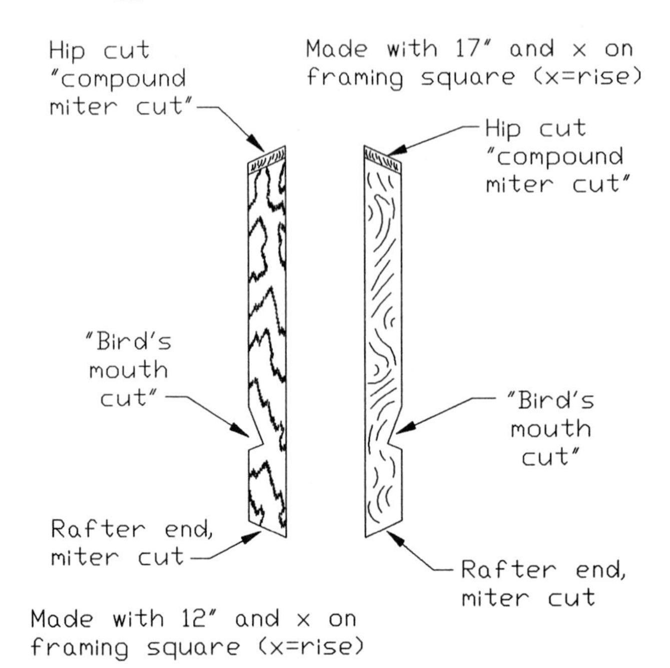

Figure 1-17 Pair of jack rafters.

Also, we must determine if the old hip was *backed* or an allowance was made in the bird's-mouth cut. One way or the other, we must lower the hip rafter to align its top edge with the roof's common rafters. Figure 1-18 also shows the backing cut or allowance for backing. *Note:* If your workers cannot perform the task of layout and cutting of rafters, you may recommend a book entitled *Framing and Finishing* 2 Ed., by Byron W. Maguire, 1989, published by Prentice Hall, Englewood Cliffs, New Jersey.

Finally, we cut two new common rafters.

With the rafters cut, we can install them in the roof. The tasks are as follows:

1. Install the hip rafter first.
2. Install the jack rafters, in pairs, beginning with the short ones first.
3. Install the two common rafters.
4. Replace the collar tie and any bracing the roof had before the damage.

Sheathing and drying-in the roof. In our project, we plan to replace the sheathing with CD plywood. The task is a simple one for carpenters. However, we must extend the plywood 1½ in. beyond the rafter ends to cover the 2 × 4's that we will install later. We always start at the lower edge of the roof with the first sheet. After nailing it in place using the schedule of 6 in. o.c. along the edge and 8 in. o.c. through the center, we snap a line and trim the excess from the hip. One-half of the hip must be exposed after the cut is made. The last pieces to install are at the ridge.

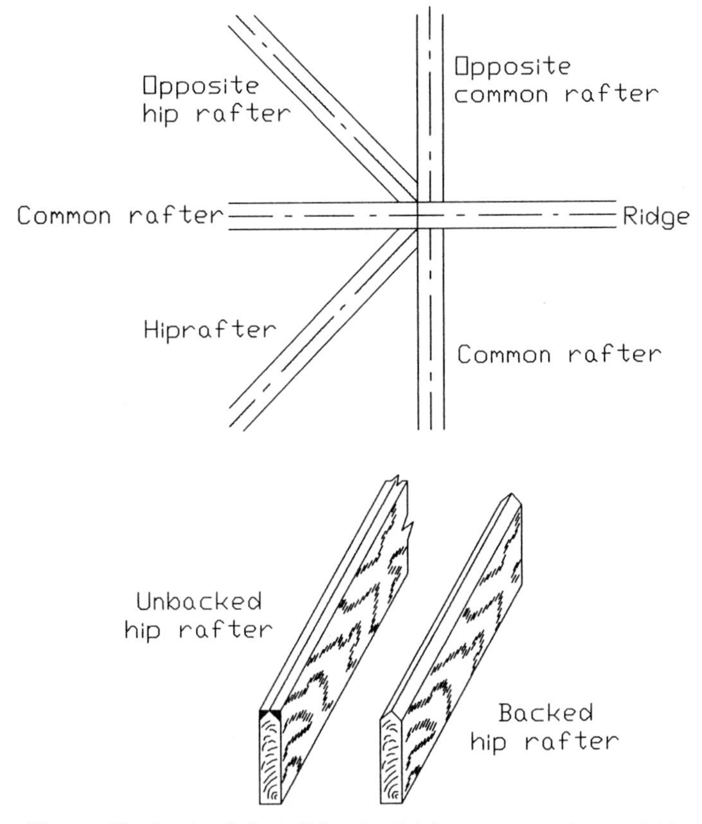

Figure 1-18 Overhead view of hip rafter joining common rafters and ridge.

We dry-in the roof with 15-lb felt, installed in the direction of rainfall. A minimum 2-in. lapover must be used from row to row. A 6-in. lap on the sides must be used. At this point we leave the shingling for later and begin working on restoring the cornice.

Restoring the cornice framing. Activity 6 directs that we reconstruct the base for the cornice. This is the close variety cornice; therefore, it has no overhanging rafter tails, soffet, and frieze board or molding. Rather, as we saw in Figure 1-15, there is a built-up wood cornice and over that base the siding contractor will mold and install paint-bonded aluminum. We have the following to do:

1. Reinstall the 2 × 4's.
2. Reinstall the molded 5/4 × 4.
3. Reinstall the 1 × 3 with bullnose (quarter-round edge).

The 2 × 4's are nailed up against the plywood and in line with the other ones already on the cornice. The alignment must be exact. We may find any possibility of spacing between 2 × 4's, from none at all to 1/2 in. or more. Because the architect desired more dimension in the aluminum, we must replace the specially cut 5/4 × 6 piece. Then the last piece is a bull-nosed 1/2 × 3 piece that extends down and over the vinyl siding. This arrangement ensures that no rain or snow can blow into and behind the siding.

Since the wood will be entirely covered, we do not need to prepaint the wood, nor need we concern ourselves with avoiding hammer marks. In addition, we can use butt cuts when we make the corner fits of the 2 × 4's and 5/4 stock. However, we will use a miter box and make the miter in the bullnose.

Replacing the aluminum cornice. We subcontracted for this part of the job since the aluminum had to be molded to fit the wood base. The contractor must first match the type and color of materials. Then he or she must mold the flat aluminum to the various bends and curves.

Some contractors use a tool called a *break* to make the bends. When they use this tool, there is a restriction on the length of each piece that can fit into the break. The other type of machine some contractors use is a molding machine where the flat metal is fed into the end and is molded into the shape. You may have seen this tool in use when continuous rain gutter is made on site.

The contractor will nail the aluminum in place with coated nails and will place a corner piece over the one corner. This piece will be made on site.

Reshingling and cleanup. All that remains to complete the restoration is to reshingle the roof according to the contract specifications and clean up the job site. Since we have already covered shingling, we will not discuss the basics. The only comment necessary is to acknowledge that a row of cap shingles must be installed over the hip after the courses have been installed on both slopes.

The cleanup activities do not begin at the end of the job. Rather, we must train our workers to make cleanup a continuous practice. Only by doing this can we minimize accidents and injury. Contractors must pay for worker's compensation, and some of us have health plans; both cost money although we might consider the expense necessary for the well-being of our crew. So what we should have been doing all along is as follows:

1. Carrying away old shingles as they were taken from the roof.

2. Pulling nails from the old sheathing before throwing it on the truck.

3. Removing nails from cornice and rafter materials as they are removed. Since these may be used for patterns and reuse, we need to stack them 8 to 10 ft back from the immediate work site.

4. Fallen nails and loose nails need to be picked up regularly to preclude punctures.

5. Trimmed ends of new rafters must be thrown on the truck or stacked away from the immediate site when all have been cut. Likewise, excess plywood sheathing must be removed from the roof and scaffold to eliminate faulty surface.

6. Excess shingle pieces must be trapped on the roof or scaffold to prevent damage to shrubs around the house and to prevent staining of sidewalks and driveways.

7. Repair of marks made by scaffolds must be made after the scaffold has been removed.

8. An electromagnet needs to be used on the grounds to pick up nails and bits of aluminum should there be any.

9. Final raking and sweeping complete the cleanup.

Concluding comments. This restoration project was very complex since the roof design was a hip roof. The job was also more difficult since the work was done on a two-story house. Working off the ground 16 to 20 ft is always more dangerous. We had to remove all the damaged materials in a special order to prevent further damage and for safety reasons. Then we had to lay out and cut new hip jack and hip rafters as well as two common rafters. We also had to "back the hip" or "increase the depth of the bird's-mouth cut" to compensate for the height of the hip. Installing the hip will be the most difficult rafter to install. The others are smaller and lighter and therefore easier to install. Rebuilding the cornice was not difficult since we merely copied the undamaged one. The siding contractor had a small and not too difficult job matching the aluminum cornice. Installing the shingles is a routine job.

CHAPTER SUMMARY

We have explored a wide variety of problems with the roof and cornice in this chapter. Every kind of situation pertaining to the roof and cornice arising from normal wear and tear to violent storm damage have been covered. All roof and cornice work is performed above ground, from ladders and scaffolds and on the roof. It is dangerous, difficult work and requires considerable skill and technical knowledge. Some homeowners can perform many of the tasks, some would be wiser to let the skilled roofer or carpenter do the work. Pricing for the jobs will be expensive even when a fair contract price is agreed to. Restoration can be an exact replacement, or the style and architecture can be maintained while substituting modern materials such as aluminum fascia, vinyl soffet and fiberglass shingles.

2

SIDING

OBJECTIVES

To solve problems with weathered siding, cleaning, and repainting.
To replace decayed wood lap siding.
To replace damaged vertical siding.
To restore damaged vinyl siding.
To restore damaged wood shingle (and shakes) siding.

OPENING COMMENTS

Siding on residential homes is made from a wide variety of natural and manufactured materials. We, as contractors, must assist the homeowner in resolving problems pertaining to his or her home's particular type of siding. We must therefore have knowledge of the materials used in the siding, as well as the methods by which they were installed and finished.

In this chapter we concern ourselves with examining some of the circumstances or conditions that would cause the homeowner to call on our skill and knowledge. If the siding is wood type, we can expect a failing or failed painting or staining system.

But there may be significantly more work if the wood beneath the finish has decayed or has become infected with decay from whatever source. If the siding is vinyl, we would expect to replace the damaged materials, and this would require either a subcontractor, a vinyl siding contractor, with access to the proper matching materials and skills needed to install it or one of our crew with equal skills. If the siding is cedar wood siding, we would expect to replace the damaged parts or maybe the entire wall covering.

When called by the owner to the job site, we must evaluate the needs of the homeowner through careful listening to his or her complaints as well as careful examination of the conditions of the siding. The homeowner expects us to restore the siding to its original condition at a reasonable cost. Most owners do not know a great deal about painting or staining systems, except what they read in magazines or listen to in advertisements, or what it takes to effect repairs, except through word of mouth from friends and workers. Therefore, we must use good questioning techniques to arrive at the root of the problem. Once we have the answers, we can make proper recommendations for the owner. Some good questions might include the following:

1. Tell me about the repairs you have already had done pertaining to this problem with the siding.

Follow-up: How long ago were the repairs made?
What type of repairs were made?
Did you or a friend do the work? Did a contractor do the work?
Do you have the bills and do they show the exact types of materials used, such as type of paint used as prime coat, surface or top coat, and so on?

2. Tell me when you first noticed the problem.

3. Tell me and show me what damage has been caused to the siding and other areas.

4. Explain, as best you can, what you want me to do.

Notice that all the questions are designed to provide the owner with a nonthreatening environment to assist us in obtaining important information. We must be careful to create this type of open dialog to avoid misunderstandings later during contract development and signing and after the job is done.

The types of projects that we will perform in this chapter include renovation and restoration of horizontal and vertical wood siding, vinyl siding, and wood shingle siding. In each project, we will look for the simplest solution first and then treat the more difficult problems later. When we replace the materials in one project, such as decayed sheathing, we will not repeat the process again, but we will make reference to it. Let's begin with the problem of badly weathered lap siding.

PROJECT 1. RESTORATION OF BADLY WEATHERED LAP SIDING

Subcategories include burning or heat ironing away old paint; pressure-washing removal of paint; removal and replacement of beveled siding; removal and replacement of drop siding; removal and replacement of 1 × 12 lap siding; corner joint treatments when replacing the damaged siding; preparation for painting the new and cleaned siding.

Primary Discussion with the Owner

Problem facing the owner. In Chapter 1 we covered the essential need for stating the work to be done in reasonable terms. This holds true for each project in this chapter. For this project, we arrive at the site and examine the wood siding. After discussion with the owner, we discover that the painting system is 5 years old and was put over the old system with no preparation before repainting except some scraping. We notice that on the south side, especially, several of the boards are split and show signs of decay or rot. The corner boards are also in poor state.

We walk around the house with the owner; we make notes of the problems and discuss these openly. From the contractor's viewpoint, we should replace the decayed and badly worn wood before repainting. We must also clean the old paint off the wood. We need to receive agreement with the owner.

Alternatives to the problem's solution. There are several solutions to the problem in this project. We could scrape the old paint or pressure wash it, put spackling in the cracks, and paint the wood with a base coat and two top coats. This solution, although the least expensive, provides an immediate solution but one that lacks quality and avoids the best or more permanent solution. The second solution is to replace only the siding pieces that have splits over 6 in. long and those that have visible rot. Then follow up with a scraping or pressure-wash job and painting system. This solution is much better, but is still not the best quality job. In this solution, the contractor is making the necessary replacement of siding before repainting. The best solution is to replace all the siding with splits and decay, burn or sandblast off the old paint, and repaint with an approved system. This combination of techniques produces the best restoration and the most long-lasting one as well. It is the one we shall perform.

Statement of work and the planning effort. Our first statement of work is that *we must restore the home to its original character and quality.* The siding that we replace must match the original exactly. This might pose a problem and affect costs. We must *paint the house the same color as it was originally.* Other work to be done includes *replace defective and damaged siding and corner boards.* We must also *sandblast or burn off the old paint prior to applying the new painting system.*

The planning effort is routine except for the matching of the old drop siding. Although it is still available in many lumberyards, the exact dimensions of the piece and coped area seldom match exactly. Therefore, we may have to have the pieces custom made or modified to match the old siding. If this is the case, we must expect a short delay before beginning actual reconstruction on the project. The carpentry parts of the project can easily be done by one carpenter. The painting system will be subcontracted to a painting contractor, and in his or her contract a specification to sandblast or burn off the old paint will be stipulated.

On the other hand, the owner may have called a painting contractor, since the predominant work to be done is the painting effort. If this is the case, the painting contractor would be expected to have the knowledge to evaluate the condition of the wood and could estimate the extent of replacement required. If there is any question or doubt, then the expertise of a carpenter would be needed to estimate the replacement of wood, and the painting contractor would include these estimates along with his or her estimates.

Contract. In Chapter 1, we defined the elements of a contract that can stand the test of legality. We are obligated to prepare one for the owner for this project, too. We shall use the bid/contract format where the bid, if accepted, is also the contract. The problem we have putting a bid together for this project is that much of the wood is covered with paint and therefore we will not be able to accurately determine the amount of wood siding that needs to be replaced.

This situation does not permit a fixed price contract, which is what the owner desires most, since that would indicate the total cost. In fairness to ourselves as contractors, we must ensure that all variable costs associated with the job are covered. In addition, we must also ensure that a percentage of fixed costs is covered and an allowance for profit is included.

We have several options.

1. We can make a dual-part contract where the painting work is provided at a fixed price and the carpentry work and materials are time and materials at estimated or specified costs. This option grants the owner a fair option. Some of the costs are known exactly, and some may be estimated by him or her. At least there is a range of dollars. With this figure, he or she can make a decision as to whether or not to proceed.

2. The bid/contract proposal could be entirely time and material. An open-ended contract of this type would certainly benefit the contractor. It would ensure that every cost would be covered; allowances for fixed costs and profit could be included as factors in either hourly rates or materials, or a part to either. The owner would have to be a risk taker to accept this type of contract; however, there would be conditions such as that the owner participates in determining whether or not siding is replaced, requires a specific number of square feet of wall surface be worked each full day at the rate agreed to, and participates in the selection of the painting system, thus affecting the materials costs. The problem for the owner is that he or she must be available throughout the project. If the owner is earning $15.00 per hour at work, he or she sacrifices $120.00 per day in economic gain. This can become prohibitive if the job lasts a week or more. Therefore, a pure time-and-material contract, even with specifications, would be an unlikely owner's choice.

In this project, we and the owner agree to option 1 with conditions. Our obligations are to prepare the body of the contract in several parts. One part will be the fixed cost to satisfy the stripping off of the old paint and repainting the entire siding, overhangs, and eaves. Part 2 of the contract must be for the time and materials to replace the decayed and split siding and corner boards.

The body of the contract could look like this:

Part 1. For the fixed price the old paint shall be stripped to bear the wood by a method of the contractor's choice. All new wood and old cleaned wood shall be primed and have two top coats of latex system applied. The project price includes cleanup of the job site and removal of overspray from windows and doors and any other surface area. The owner shall have choice of color, with a limit of a primary choice and one alternate color. The quality of the paint system will be expected to have the manufacturer's life of 10 years. The contractor shall have a workmanship warranty of one year for defective workmanship to include poor preparation and application. Lastly, all surface marks such as nailheads, knots, small cracks, and gaps between corner boards shall be filled or sealed appropriately prior to painting.

Total Fixed Cost $(Estimate 80% labor + 20% materials)

Note: Built into the labor costs are fixed and variable costs for labor-associated support items such as worker hourly wages, owner's salary, allowance for profit, insurances, taxes, FICA, unemployment tax, office help, and others. Built into material costs are transportation costs, allowance for waste, costs for tools and equipment replacement, and travel costs, to name the most predominant ones.

Part 2. For the repair and replacement of decayed or split siding and corner boards, a time and materials contract shall be in effect. Boards with decay through the board shall be replaced. Boards with splits longer than 12 in. and ends with splits longer than 6 in. will be replaced in part or in whole with matching pieces. Nails will be countersunk and filled; other cracks, holes, and splits will also be filled and sanded smooth. Proper care must be used to ensure that subsurfaces beneath the siding are not left to the weather. If sheathing is also damaged, its repair is authorized by this contract with prior approval of the owner. The following stipulation must be enforced: When the paint has been stripped from the walls and the siding is exposed, this part of the contract shall be further defined as to costs, based on the costs per hour estimated and material costs, before work begins.

Labor costs: $28.00 per hour per carpenter

Material costs: Estimate $3.25 per lineal foot of siding and $2.25 per lineal foot of corner boards. Prices subject to variation of availability and shipping costs.

Due to the variables in this contract, we would probably want a deposit. Some contractors need the cash for supplies. Others prefer to have the funds in a separate bank account, which ensures its availability when certain aspects of the job are finished. Some schedule of payment may be included in this contract. The timing of payment varies among jobs, the value of the job, and the trust the two parties have in each other. For this contract, due to its short time frame for completion and modest outlay of cash, no deposits are required, but payment in full is expected within 10 days after job completion. So we can add the following:

No deposit is required at the bid/contract signing. Payment in full is required within 10 days after completion of all work and clean up.

Material Assessment

Direct materials	Uses/purposes
Painting system	Prime and top coats
Putty or spackling	Fill holes and cracks
Siding	Replacement boards
Corner boards	Replacements
Nails	Fasten siding and corner boards
15-lb Felt	Installed behind siding and corners
Caulking	As required

Indirect materials	Uses/purposes
Sandpaper	Smoothing
Plastic sheets	Protection of surfaces not to be painted
Masking tape	As required
Sand	If sand blasting
Detergent	If pressure washed

Support materials	Uses/purposes
Ladders	Access to work surfaces
Scaffolding and planks	Work surfaces
Hot-iron paint remover	Burn off the paint
Sandblasting machine	Clean walls of old paint (rental if used)
Power washer	Washes away old paint (rental if used)
Rollers, brushes, etc.	Apply paint
Spray outfit or power roller	Apply paint (alternatives)
Carpentry tools	Reconstruction
Power handsaw	Eases cutting siding and corner boards
Extension cords	Provide electricity from house to work site

Outside contractor support	Uses/purposes
Painting contractor or	Subcontractor for painting
Carpenter contractor	Subcontractor for replacement of siding and corner boards

Activities Planning Chart

Activities	1	2	3	4	5	6	7
1. Contract preparation	×						
2. Scheduling and materials	×						
3. Paint removal		×	×				

Time line (days)

(continued)

Activities	Time line (days) 1	2	3	4	5	6	7
4. Siding removal and installation			×	×	×		
5. Sealing and priming				×	×	×	
6. Painting system (top coats)					×	×	×

Reconstruction

We have examined the various aspects of preparing the contracts and the various necessary discussions we must have with the owner. Therefore, we will not retrace the information already covered. But we need to discuss several aspects about activity 2.

Scheduling and materials gathering. These activities are two interrelated aspects. To properly schedule the entire project, we must understand several conditions that we face in this generic project. While at the job site in discussion with the owner, we observed the type and dimension of the siding and corner boards. With this information, we determined easily that corner boards can be prepared on site from No. 2 common pine or fir. However, the replacement of the siding is another matter. Because it is the lap type, we can make a very close approximation, but maybe not accurately measure its full dimensions without prying some loose. Furthermore, we are not exactly sure that any needs replacement until we remove the old paint. So our scheduling may require (1) sandblasting or burning off the old paint by the painting contractor, (2) a delay to acquire the siding and other materials needed, and (3) rescheduling the painting contractor after the carpenters are done with the replacement effort.

Notice that the activity chart does not indicate a delay between activities. There may not be any if, for example, 1 × 12 siding is the type used. If there is no delay for buying the materials and good weather prevails, we should complete the actual renovation in 6 days.

Buying the materials is not a difficult task, but several supply houses will likely be used. The lumberyard or builder's supply should provide the materials needed by the carpenters. The paint and paint support materials can be bought in a specialty shop such as a Sherwin Williams store. The rental equipment needs to be scheduled and picked up the day needed to minimize costs. Each of these materials and rentals requires personnel and transportation, which are indirect costs usually attributable to variable costs. Several hours will be used to pick up materials and transport them to the job site.

Paint removal. In our generic project, we contracted to either sandblast or burn off the old paint. In this chapter we will provide a short explanation of the three methods for paint removal, which are burning it off, power washing it off, and, finally, sandblasting it off.

Burning off the Old Paint. The technique of burning off the old paint is one of the earliest methods. In the first half of this century, the painter would use a blow torch to scorch the paint and with a wide-blade putty knife scrape the bubbled-up paint. The trick or skill was to apply sufficient heat and flame to scorch the paint and free it from the wood underneath. This technique worked very well. Seldom if ever did the painter cause a fire. The exposed wood was free of most of the paint, since the wood grain was readily seen. After the scraping, the painter sanded the surface slightly to remove any residue of paint and then applied the new paint.

If we use the technique today, we use a butane torch on the end of a long hose, which is attached to a 15-lb or larger tank full of gas. The technique of applying heat remains the same. We only apply enough to bubble the paint, but not so much to cause flaming or burning of the wood beneath. The use of the long hose and pressure tank simplifies the process somewhat, since we do not need to repeatedly pump the handle on the blow torch. And the weight of the new torch is much less than the blow torch.

An alternative to burning that is safer in many respects is scorching the paint with a heating iron. Heat directed at the paint scorches the paint and we then scrape it off. For this technique we require ladders and scaffolds.

Pressure Washing off the Old Paint. The principle behind this technique is the force of water under high pressure. A tank with a pressure-producing pump is operated by either electricity or gasoline. The pressure forces water from the nozzle in a high-velocity stream that blasts against the painted surface. Where the paint is already flaking or loosened from the wood, the water washes it free and away from the wood. This technique is very fast and relatively safe. The problem is that paint that is still securely bonded to the wood is left unaffected. This means that the newly applied paint will not be universally smooth, but may very well show some variations. It is possible that we can do this task from the ground without ladders.

Sandblasting off the Old Paint. Both systems for paint removal discussed above do not affect any surrounding surfaces and, therefore, simple protection schemes such as covering the ground and shrubs are all that is required. When using the sandblasting technique, glass and surfaces that will not be painted must be protected from the high-velocity sand. In addition, we will need to have protective face masks to filter out the sand and protect the eyes as well. We will also require some scaffolding to ease the work.

Sandblasting equipment will probably be rented, and the renter will also have the special sand needed. The system is electrically operated and the sand, under high pressure, is ejected through the nozzle on the end of the hose. As the sand strikes the paint, it breaks up the paint, and the paint particles fall to the ground. Because of the sand stream and an adjustable pattern, corners and flat surfaces can be quickly cleaned.

After the paint has been removed, the area surrounding and beneath the walls should be cleaned up. This simple act will eliminate a situation that can be a problem if left till the end of the job.

Siding removal and installation. With the paint removed, we can accurately determine how much of the siding needs replacement. In this section of our generic project, we look at several different types of siding and briefly identify how to remove and replace the pieces. Figure 2-1 shows three kinds of horizontal siding. In Figure 2-1a, we see novelty siding, which is still in use today. In Figure 2-1b, we see clapboard siding, also called beveled siding. In Figure 2-1c, we see colonial siding; this actually is a rectangular board or composition board. Of the three, the one most commonly employing corner boards is novelty siding. The other two types normally use an overlap joint at the outside corners; however, there were many jobs where a 45° miter or overlap butt joint was used.

Our problem is to remove the split or decayed pieces and replace them with new materials. This is where the problems begin. Let's take each separately.

Novelty Siding Removal and Replacement. As a rule, the corner boards and window trim were nailed over the top of the siding as Figure 2-2 suggests. Thus we must first remove these pieces to get to the siding underneath. Since the siding was nailed with either 6d or 8d common nails and countersunk, the likelihood of removing the pieces without splitting or other damage is slim, so we can expect to replace them.

We first remove a piece and use it as a sample when we buy the replacement siding. What we are likely to find is that no exact replacement is available. Each mill run of the siding should be exactly the same but usually is not. Therefore, we may need to have a mill modify some that is available or just custom mill the entire lot. A shop with a shaper can do the job easily.

Notice that this type of siding makes 100% contact with the sheathing. There should be a layer of felt or other type of paper covering the sheathing. If decay or water penetrated the siding, it is likely that the paper prevented damage to the sheathing. If this is true for your project, the siding can be cut and installed without concern for the sheathing. Joints along a row usually butt with a perfect fit. They may also be mitered, but seldom are.

Once the siding pieces have been replaced, we make new corner boards and window and door trim as needed to complete the restoration carpentry effort.

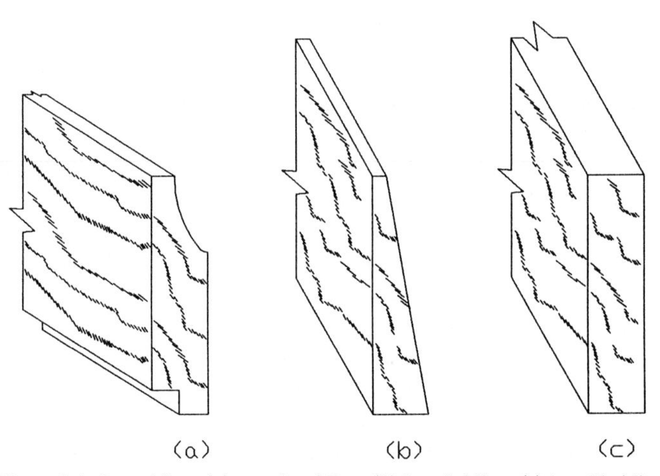

Figure 2-1 Lap siding: (a) novelty siding, (b) bevel siding, (c) 1 × 12 siding.

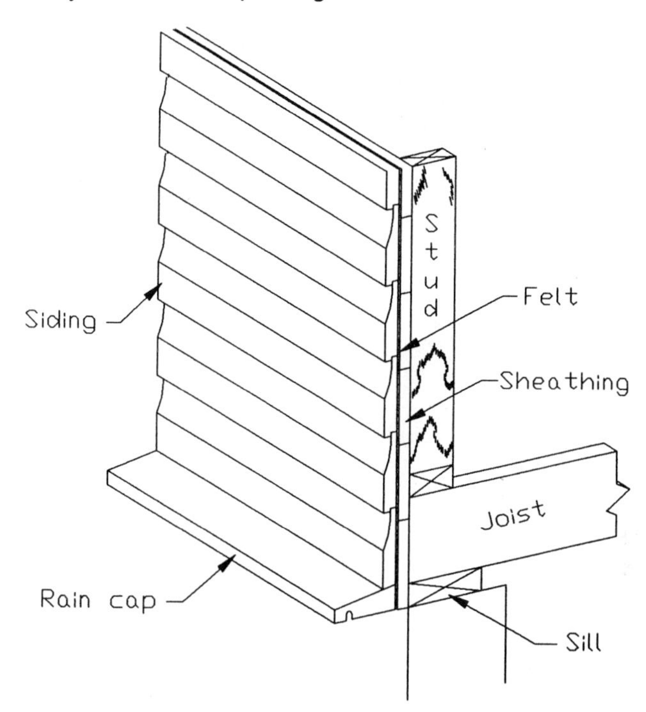

Figure 2-2 Novelty siding installation.

Clapboard Beveled Siding Removal and Replacement. Several different sizes of clapboard siding were made and are still in use. They are the 6, 7, 8, and 10 in. When the 6-in. size was used, the exposed surface was usually about 5 in. When the 8-in. siding was used, the exposed surface was about 7 in. Nowadays, this type of fir siding is paper wrapped at the factory and sold in square-foot lots per bundle. Masonite and other corporations make a fiberboard beveled siding that has the same dimensions as the wood variety. It is more impervious to decay and rot. Figure 2-3 shows how this siding looks when installed. Notice that the siding does not fully touch the sheathing. Also be aware that special methods are used for interior corners and exterior or outside corners and to start the first row; these details are shown in the figure as well. We also must mention that on very old homes that employed clapboard the siding was nailed directly to the studs without first adding sheathing. However, as insulation became important, about the end of World War II, fiberboard impregnated with asphalt was used under the siding.

We must understand that when we replace the siding, the first row has a starter and not a full board. The starter must be styled to ensure that each overlap closes perfectly when nailed and that no stress is placed in the middle of the piece being nailed. This detail is also shown in Figure 2-3.

To remove the siding, we insert a flat chisel under the overlap near a nail and pry it loose. We must start with the last row to be removed. To avoid splitting the row we want to leave on the wall, we locate the face nails and drive them through these boards. This makes holes that can be filled later, but it prevents splitting the boards.

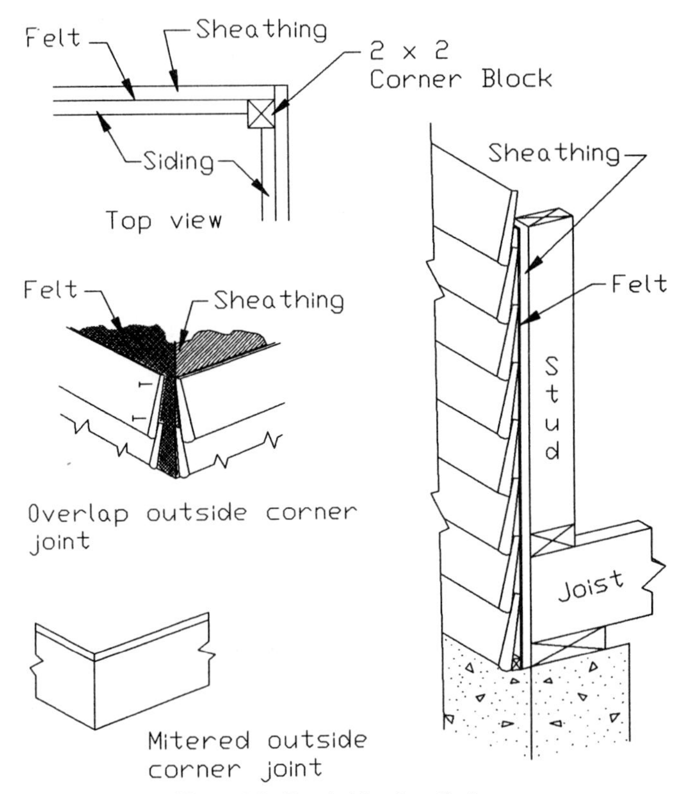

Figure 2-3 Bevel siding installation.

Moving down one row, we pry the board free and remove it. After all boards are removed, we begin at the bottom of the wall and work our way up a row at a time, installing the new siding.

Colonial Siding Removal and Replacement. Colonial siding was extremely popular after World War II. It was easy to install and looked really good. Various board widths were used; they included 1 × 8, 1 × 10, and the most common, 1 × 12. As Figure 2-4 shows, this siding also required the use of a starter, but the starter never showed, whereas it sometimes showed in beveled-siding application. Two types of exterior corner treatments were generally used that did not employ the corner board. These are the same as we saw in Figure 2-3, the overlap and the miter. Although the miter joints looked really good when done by a skilled carpenter, they were subject to opening, primarily due to the drawing power of the sun and swelling and shrinking from weather. The overlap joint was much better and stayed tight much longer. The lap joint was simple to make, and we frequently dressed the raw end wood with our block planes. The painter forces a thin coat of white putty into the grain prior to painting, and when the entire paint job was done, the corner looked perfect.

To solve the problem of fitting the beveled boards in the interior corners, we frequently installed a dressed 2 × 2. Then we just butt joined the siding to the 2 × 2.

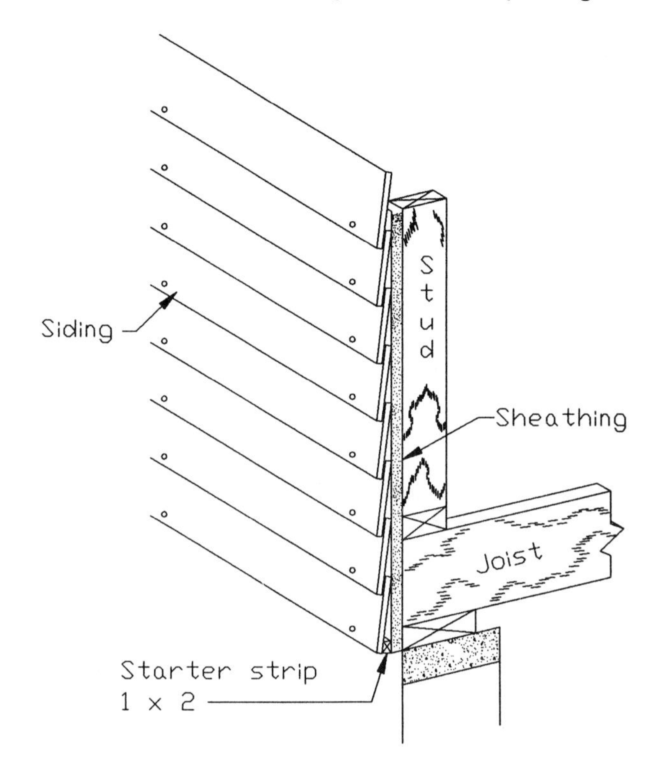

Figure 2-4 1 × 12 Siding installation.

This detail was also shown in Figure 2-3. Because of the quality of the joint, no caulking was ever required. To further prevent leaks, we installed a 12-in.-wide strip of 15-lb. felt behind the 2 × 2.

To remove and replace this siding, we start at the row above the highest piece on the wall we want to remove. As a start, we drive the nails through and pry the piece up slightly. Then we insert a flat bar between the boards on the next row down and begin removing each row that must be replaced. The replacement is opposite to the removal. The final board slides up under the row left on the wall and we nail through the top piece and fill all holes.

Sealing and priming an application of the two top coats. The raw wood needs to be primed, and because of our skill the only filling to be done is the nail holes and end cut on the siding. For this job we do not require any caulking. The prime coat for this job calls for latex, which we apply with a spray outfit or roller. After it dries, we apply two top coats of exterior latex at 1.5 mils thick. According to recommendations from the paint manufacturers, we achieve the best life expectancy by applying the two specific coats versus heavier coats or thinner coats.

Concluding comments. All lap siding should be replaced according to the descriptions provided above. These may include composition boards, asbestos-cement siding, and others. The principles even apply to replacing vinyl siding, but there are several variations. Therefore, we will examine vinyl siding separately.

PROJECT 2. RENOVATION OF VERTICAL BOARD AND BATTEN SIDING

Subcategories include removal and replacement of reverse board and batten panels; spray staining siding; replacing drip caps.

Primary Discussion with the Owner

Problems facing the owner. When the contractor is called to the job site by the owner, who is concerned with badly weathered or bug infested, rotten boards and split boards, the contractor must recognize whether or not there really is a problem. In many cases where cedar or cyprus is the siding material, water from rain will not penetrate through the siding because the wood immediately expands as water hits its cells; therefore, small splits do not cause leaks. These woods react just like wood shingles on a roof. Figure 2-5 shows a board and batten siding installation. Notice that the boards do not need to touch. In fact they are usually spaced 1/4 to 3/4 in. apart. The batten, a 1 × 2, is nailed over the gap.

The contractor must evaluate the conditions of the siding. First, the end wood near or touching the ground or built-up garden material is prone to infestation and rot. The blackened wood color indicates rot and fungus. While the owner is present, the contractor should make several probes with a pencil or small screwdriver to reveal the type of damage and its extent. It may turn out to be dry rot or, if little white insects are present, it could be termites. However, other bugs may also be present, such as carpenter ants.

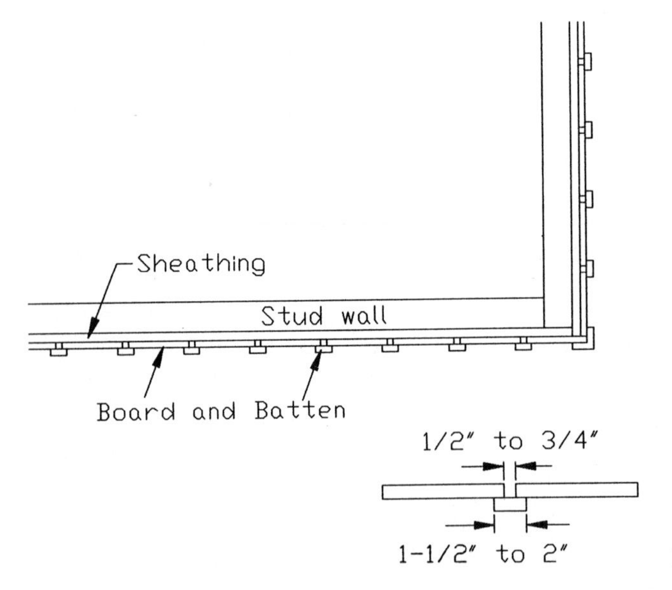

Figure 2-5 Vertical board and batten.

How the condition happened depends largely on the surrounding conditions. We, as contractors, must explain that vegetation generally traps moisture even when the wind blows. The open-end wood cells attract and absorb the water and make a perfect housing condition for insects and fungus. Because the wood can not dry, it rots, fungus grows, and wood is eaten or eroded away.

We also discuss warped boards, splits that are extensive, and the battens that have been pulled away by the sun's pulling power.

In our discussion with the owner, we explain that we will replace the boards as full-length pieces versus just replacing the damaged portions. This is in keeping with the original concepts when applying the vertical siding. The board reaches from just below the foundation sill to the soffet, frieze, or overhang. On the gable end, we usually use a rain-drip molding between the lower wall pieces and gable end pieces. This detail is shown in Figure 2-6.

The job is not difficult. The difficulty is in matching the materials and aging the color. If the material is cedar, it is generally available, but if it is cyprus, it may have to be shipped in from southern states dealers. Treated lumber should not be used since it cannot be matched for grain, texture, and color. Not much pine was used for this type of siding.

Note: Composition board made by several firms in the reverse board and batten style is prepainted and, therefore, is not a problem to match for style and color.

Alternatives to the problem's solution. There are very few options if the damage includes any of the causes discussed with the owner. We must remove the damaged materials and replace them with like materials. We may have to alter the

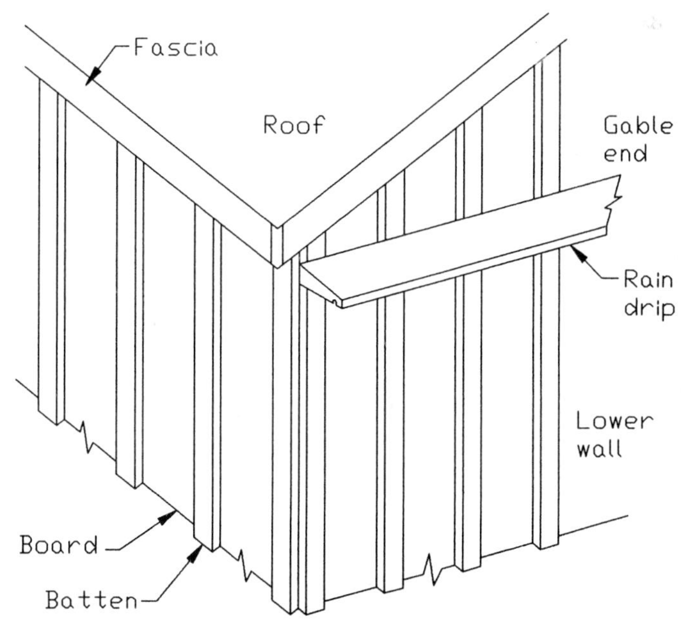

Figure 2-6 Rain cap between lower wall and gable end in board and batten installation.

dimensions of the new wood and cut battens to match because lumber has been downsized in more recent times. We may have to remove trim around the windows or, at the least, loosen the various pieces to insert new felt behind the trim. We may have to replace the copper drip cap (if it is damaged) or galvanized rain cap that was installed over the window and door trim if it is rusted through. Again, we must discuss the possibility of further damage to sheathing or framing that will be revealed when the siding is removed. However, we should make a rudimentary examination while on site by removing one piece of siding and two battens. If the siding has been stained in the past, we should recommend restaining the entire side of the house versus just the new boards. If the siding is left natural, we must age the boards and battens to approximate the color of the siding.

Statements of work and the planning effort. The major effort is to *replace all damaged vertical boards and battens agreed to between owner and contractor.* Furthermore, we must ensure that *the finished color matches the original color by restaining the entire house or aging the wood.* We also need to *ensure a waterproof job and clean up the job site when done.*

For our generic project, we will be replacing 50% of the siding on the weather-beaten north side and 30% fungus-damaged boards on the south side. We will have to cut around windows and doors, but there are no gable ends to work on. The house was stained with a heavy-bodied stain 3 years ago, so we should not have a problem matching the stain. However, we plan to restain the entire house after discussion with the owner. We will plan for two coats of stain applied with a spray outfit.

The carpentry work will require either one person for a planned week or two carpenters for two days. Acquisition of materials will take one day. A subcontractor will be used for staining the building.

Since this is a small job and the owner is financially able to pay, we will not require a deposit. However, payment in full is expected on job completion.

The type of contract we will offer is a *fixed price* contract.

Contract. As stated above, we employ the fixed-price contract for this job. However, we will include a simple clause to cover labor and materials should we uncover damage to sheathing, windows, doors, or framing. Our office will prepare a fixed-price subcontract for the painter and provide a window (range of dates) for completing the work.

Materials Assessment

Direct materials	Uses/purposes
Siding and battens	Replace damaged siding
Stain	Preserve and color the siding
Nails	Fasten boards
Putty	Fill nail holes (optional)
Aluminum flashing	Replace rain-drip flashings
15-lb Felt	Waterproof under siding and around windows and doors

Indirect materials	Uses/purposes
Masking tape and plastic	Protect surfaces not being stained
Mineral spirits	Clean spray outfit and brushes

Support materials	Uses/purposes
Extension and step ladder	Work surface
Portable workbench	Work surface
Power saw	Ease cutting
Carpentry tools	Construction

Outside contractor support	Uses/purposes
Painting contractor	Stain the house

Activities Planning Chart

Activities	Time line (days) 1	2	3	4	5	6	7
1. Contract preparation	X						
2. Scheduling and materials	X						
3. Remove and replace siding		X	X				
4. Staining the siding				X	X		

Reconstruction

Contract preparation. The contract preparation and final negotiations and signing can be done in less than a day. However, the painter must be given the statement of work pertaining to the job, the location of the job, and time to assess the work, and we must cost the job for direct and indirect materials as well as fixed and variable operating costs. These activities are normal and must be explained to the owner.

The clause pertaining to unforeseen repair work behind the siding needs to be inserted, and the owner must be reminded, at signing, that it represents time and material costs.

Scheduling and materials. In the section above on planning, we discussed the need to locate suitable materials that match the siding on the house. Several phone calls from the office should quickly determine whether or not the siding is locally available in the sizes required. The painting contractor should also do some quick checking to find suitable stain. The nails we use must be galvanized or coated to prevent rust stains on the face of the wood.

The schedule for picking up materials for the carpenters is our responsibility and should be done on the morning the workers leave for the job unless a better solution is available. The carpenters need to take the power saw and other company tools with them, but, as usual, they use their own hand tools. The carpenters must know that the job is scheduled for 2 days unless they find damaged materials behind the siding. Then they must be informed about the need for instructions to proceed beyond the original scheduled work.

The painter expects to receive a call as soon as the schedule is set. He or she can react to the information and set up a time frame with the owner to spray the entire house. Since there will not be any damage to the siding if left unstained, the painter's schedule can be somewhat flexible.

Removing and replacing the materials. Whereas we had to be extremely careful to avoid damaging lap siding in the previous repair job, we do not have that same condition in this project. The removal tasks are as follows:

1. Remove the battens covering the joints where the boards are to be removed.
2. Remove the frieze molding or trim covering the upper end of the siding.
3. Remove the damaged boards and any trim around doors and windows that overlap the siding.
4. Examine the flashing over doors and windows for damage, if it exists. Remove if necessary.
5. Examine the wall behind the siding by slicing the felt.

Then we replace the siding in the reverse order. With the siding removed, we can determine how the siding was nailed. For example, the wall could have been fully sheathed with 1 × 6 tongue and groove, plywood sheets, plywood sheets on the corners, and fiberboard or insulation boards elsewhere; no sheathing but 2 × 4's cut in between the studs under the felt; or in some very early 1900 or late 1800 houses, some unusual arrangement. With this information we can adopt a nailing preference and determine how long the nails must be.

Next, we must dry-in the wall with new felt. After that, we begin the installation of new siding boards first and the battens last. As we nail each batten, we must countersink the nail heads. Next we reinstall the frieze boards or trim. Our final task is to clean up the job site.

Staining the siding. The last activity is application of the two coats of stain by the subcontractor. The contractor should bring ladders and a spray outfit to the job site along with other materials, such as drop cloths to cover shrubs and areas that will not be stained. Heavy-bodied stains cover the wood readily. The rough texture of the lumber decreases the square-foot coverage per gallon to approximately one-half the amount required for planed wood. This translates to about 100 to 200 sq. ft per gallon. The application of stain on siding does not require wiping. Wiping stains are used for furniture and cabinets. Acrylic latex and alkyd vehicle stains use oxides for pigments and are full covering. Oil stains, seldom used on exterior siding, use natural colors and are semitransparent. Some stains also contain varnish or plastic sealants to help extend the life of the finish.

Concluding comments. The project of restoring the siding where vertical siding is the material is not very difficult. Yet, we must emphasize that the effort requires careful planning, quality workmanship, and quality materials and their application. Furthermore, we must cover all variable costs and apportion the fixed costs according to our company's yearly plan.

PROJECT 3. REPLACING DAMAGED VINYL SIDING

Subcategories include replacing vinyl vertical siding; replacing vinyl horizontal lap siding; caulking; insulating behind the siding; replacing aluminum siding.

Primary Discussion with the Owner

Problems facing the owner. The contractor is faced with several problems that must be dealt with while making the examination of the house with the owner. These include the method used to install the siding in the first place, the extent of the damage, the likelihood of obtaining exact replacement in both style and color, and having the proper work force to do the work and the base behind the siding. Since the vinyl is damaged, the contractor is able to see behind the siding and determine the base. This also adds vital information that will affect the cost of restoration.

Although most general contractors have a good understanding about vinyl siding, they frequently use the services of a specialist, rather than their own carpenters. If this is the situation, when dealing with the owner, we should bring the specialist along or have him or her meet with the owner. If, for example, I were called by an owner, I would meet with the owner and bring a specialist to ensure the proper interface between the owner and myself. My company is at stake.

We must also be aware that many of vinyl companies warranty their products for 50 years. Therefore, the owner's insurance agent and possibly the vinyl company agent should meet us there to discuss the liabilities and insurance coverages. When insurance companies are involved, there should not be any delay, but there could be. In every case, we should make temporary repairs to prevent further damage.

For this problem, we try to obtain details about the siding from the owner. This information helps identify the original manufacturer, the age of the siding, the style and color, and possibly the installation contractor. With this information, we could talk directly to the installer and deal with the local manufacturer's representative. Some questions that are important to both the owner and us are the following:

1. Is the vinyl style and color still available?

2. How long will it take to obtain replacement pieces and trims?

3. If the siding is not made anymore, how long will it take to have a special run to match the house?

Alternatives to the problem's solution. There are really no alternatives that would satisfy the owner. We must find an exact replacement. To use an alternate product whose pattern, simulated grain, and color only approximate the original would lower the value of the house. Because of the limited options, we cannot even provide any idea of estimated or actual costs; thus we must make arrangements to meet again with the owner.

Statement of work and the planning effort. The project we are faced with require us to *remove and replace all damaged vinyl siding and trim.* Furthermore, we must *match the original siding and trim for size, color and texture.* We also need to *use warranted materials and quality installation practices.*

For the planning effort, we may have a real problem obtaining the exact replacement materials. Our plan could take two approaches. We could have one of the craftspeople or our salesperson perform the research by starting with phone calls with information obtained from the owner. Locating the materials is the most important issue at this stage of the planning effort. The alternative to using our personnel is to subcontract the job to a specialist. If we opt for this approach, we must make an appointment for the siding installer to come to the office and discuss the project, as well as take him or her to the job site. Since the specialist is in the business, he or she will have more resources than we do.

After locating the materials or having to wait for a special run from the factory, we can plan the rest of the job and begin to construct the contract. For planning purposes, we would use the subcontractor. So for this discussion we assume the work described in the remainder of this section is about the contractor's and subcontractor's work without distinction. Materials must be ordered and picked up. Damaged siding and trim must be removed from the walls and replaced with new materials. The job site must be cleaned after all work is done.

Contract. After locating the siding, we are able to provide a fixed-price contract to the owner. Since we are the general contractor, we only bill according to time investment attributable to our company. To this pricing, we add the subcontractor's price for the work. In reality, the owner is not being overcharged as he or she might suspect. Notice that we are not planning to mark up the subcontractor's price to make profit that way. Our legitimate approach does not prevent charging for variable and indirect costs, as well as direct and fixed costs plus a percentage for profit. The body of the contract would stipulate:

> **Description of work:** For the sum shown below, we agree to remove the damaged vinyl siding and trim as determined on site (approximately ××× square feet) and replace it with exact matching materials. This pricing includes cleanup of the site after the work has been completed.
>
> No specific start date has been set due to the need to import the replacement vinyl from the factory. However, weather permitting, the complete on-site work should not take more then 3 days.
>
> No prepayment is required, but payment in full is required within 7 days after completion of satisfactory work.

Material Assessment

Direct materials	Uses/purposes
Vinyl siding	Replacement materials
Vinyl trim	Replacement materials
Nails	Apply vinyl
Starter strip	Replacement materials
Caulking	Sealant if required

Indirect materials	Uses/purposes
15-lb Felt	Underlayment if required
Rake and trash bags	Cleanup

Support materials	Uses/purposes
Ladders	Construction platform
Workbench	Eases cutting materials
Carpenter tools	Construction

Outside contractor support	Uses/purposes
Siding contractor	Perform the work of installation

Activities Planning Chart

Activities	Time line (days) 1	2	3	4	5	6	7
1. Contract preparation	X						
2. Scheduling and materials		X					
3. Vinyl removal and reinstallation			X	X	X		
4. Cleanup					X		

Reconstruction

Contractor support. We have examined most of the contractor's involvement with the owner and subcontractor. However, we must mention that as prime contractor we are obligated to ensure several aspects. One of these is that the subcontractor perform the work and provide the quality we fully expect. Another aspect is the subcontractor's timeliness. We will schedule the job in agreement with the subcontractor and the arrival of materials. He or she therefore is expected to arrive on the job site on time and prepared for work. We also must plan to meet with the owner after the work is done and make the final inspection. Also, we must be available to either the owner or subcontractor or both as called on.

Contract preparation. We allow one day for the actual preparation of the contract. This should be adequate since we only have to calculate our expenses and include the subcontractor's price. We will open an account in our computer for accountability of income and expense and payment to vendors for materials and to the subcontractor.

Materials and scheduling. Looking at the action plan, we are not able to see the delay between signing the contract and the ordering and acquisition of materials. In a more elaborate chart, which we might have in our computer or on a wall, we would show the time delay with specific dates. Then, when the materials arrive, we would contact the sub and owner and provide the dates when work would be performed. These are days 3, 4, and 5 as shown in the action plan.

Vinyl removal and reinstallation. Figure 2-7 illustrates the principles of vinyl siding installation and we shall refer to this figure several times. The removal of the siding is quite simple, since each piece snaps into place at the bottom. The curved end is inserted into the retaining track of the previously installed piece. No nails are used at the lower end. However, nails are installed at the slotted ridge along the top. Carefully note that the nails are not driven home as one might do to a piece of plywood. Rather, they are driven in until they are snugly against the vinyl but do not bend it. This ensures a sound anchor and a smooth even track for the next piece of siding.

The contractor in our generic project was able to reuse some corner pieces and end pieces, as well as those around the window and door. These different pieces are shown in Figure 2-7. These pieces are channels that we slide the siding into and, from all appearance, make one wonder about water tightness. But there will be no leakage and the channels permit full water runoff. Joints in the run of the siding are usually simple overlaps of 3 to 4 in. In some installations, the contractor slips a thin, stiff piece of insulation behind the siding, which aids both in insulation and in maintaining the shape of the siding.

Care must be used when ladders are placed against a wall covered with siding. We usually pad the tops of the ladder to prevent scratching, punctures, and scarring.

Cleanup. Our cleanup on this job is very simple and quick. The vinyl is easy to spot and pick up. Our most serious challenge is picking up old nails that were removed and new ones that were dropped. For this we can use a rake, but an electromagnet is better and quicker.

Concluding comments. Installing vinyl siding is a special skill and requires full understanding. Many builder's supply houses offer the products, and manufacturer's brochures indicate that the unskilled can install the siding with no problems. This is not so, and it is far better and more frequently cheaper to employ the services of a qualified contractor. A novice can even get into trouble making repairs. Notice how we emphasized the need to obtain an exact match. Any old replacement will not work. Furthermore, much of the work must be done from a ladder or scaffold, and when using these there is always the risk of personal injury.

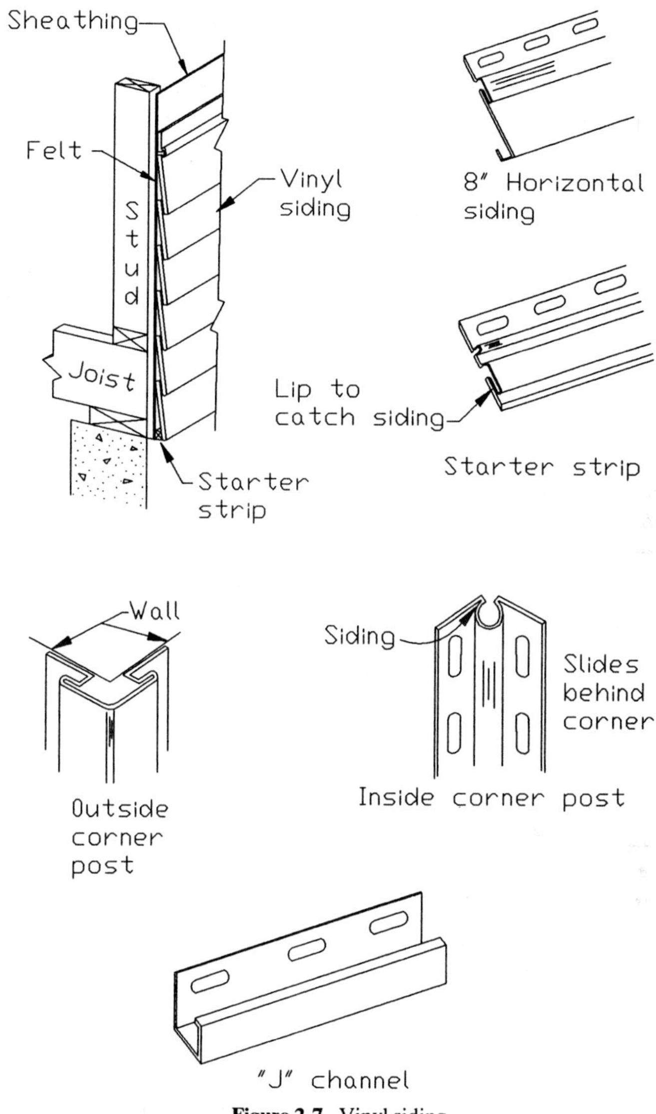

Figure 2-7 Vinyl siding.

PROJECT 4. REPLACING WOOD SHINGLES ON THE WALL

Subcategories include naturally aged, painted, and stained shingles; double-layered shingles such as sawn shingles; hand-split shakes; machine-grooved shakes, some prestained at the factory as were installed in the 1950s and early 1960s; random installation patterns, 5-, 6-, 8-, or 10-in. exposure.

Primary Discussion with the Owner

Problems facing the owner. The use of wood cedar shingles or shakes on the side walls of a house may be found on 100-year-old homes and on the newest ones constructed today. What this means in terms of dealing with the owner of a home sided in with either of these materials is that the age of the home has a direct bearing on the restoration efforts we could be faced with.

Basic wall construction differs dramatically between 100 years ago and now. In the earlier days, lattice strips of rough 1 × 3 were used horizontally across studs. Later construction included 1 × 6 sheathing boards installed on the bias. Today, insulation board and corner stiffening plywood are the base for shingle siding.

What we, as the contractor, must identify is the age of the house and its basic wall construction. We must further ascertain if the siding is the original wall covering or if it was installed over another type of siding in an earlier attempt to remodel the house. Open dialog with the owner may provide us the answers. However, we may need to make a further examination if any doubt exists.

Observation of the window units will identify each type. For example, if the house has double-hung windows with rope and pulley, we know that the siding overlaps the casing, as shown in Figure 2-8. This casing was flush with sheathing or strips. Notice that the molding is held back from the casing edge to permit the siding to overlap. In contrast, in modern treatments of the double- or triple-insulated window unit, a manufacturer's flange is used in the installation. This flange aids in installing the window unit and provides a leakproof overlap for the siding.

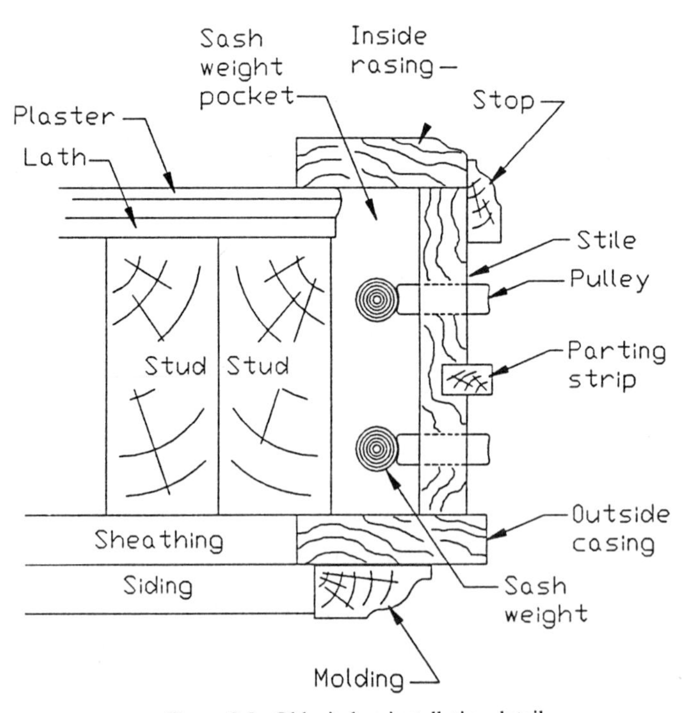

Figure 2-8 Old window installation detail.

The same treatment existed with the old-time door units as with old-time windows, and in some cases it still exists today as well. However, today the assembled door unit usually has a door trim or casing that extends over the wall's sheathing. Many carpenters install strips of 15-lb felt on the opening just prior to setting the door unit in place and nailing it. This paper provides a leakproof surface to which we can apply the shingles. To repeat, we must determine the type of door and window unit to better understand the installation of the shingles.

Next we need to discuss with the owner the type of siding that is on the house. There are three modern types and these also represent the older styles as well. They are (1) sawn shingles, (2) hand-split shakes, and (3) machine-grooved shakes. The titles or names given to each depict the technique of their manufacture. The sawn ones may be sawn with a circle saw or bandsaw, which is readily seen by observing the back surface texture of the shingle. Hand-split shingles are also easily distinguished by their random thickness and uneven texture. Some of each shingle will be quite thick while the next 1/2 in. may be thinner, although the split was made parallel, approximating a 3/8- or 1/2-in.-thick butt end. Finally, the machine-grooved shingle or shake has machine grooves that are uniform and parallel to the side of the shingle.

The owner may have papers certifying the type and life from the manufacturer. This information will also help to determine if the replacements are under warranty and should help us locate exact replacements.

The next discussion with the owner focuses on the extent of damage. Here we use the usual observation techniques. We look for blown-off shingles, decayed and rotten shingles, shingles with infestation, splits that are overlapping the lower layer, and badly stained shingles. We need to point out these observations to the owner and carefully note his or her response. The owner may wish to conserve costs and thus will opt for allowing some damaged shingles to remain on the wall. However, we are obligated to discuss the cost of a total wall replacement versus the replacement of extensive patches.

The technique of installation is next. Several techniques are used to install the shingles and shakes. One is in even horizontal rows just like lap siding. These rows are a specific number of inches apart. The common ones are 5, 6, 8, and 10 in. The random technique requires that no two shingles or shakes be even. However, there is a general row to row separation such as the four mentioned. In a further variation, each row is both random from shingle to shingle and from row to row. Each technique requires a different type of skill and time for installation. The shortest installation time is for the straight line, increased time is needed for the random installation, and the greatest time is needed for the double random technique. In addition, there may be two layers per course, with the underlayer made from thinner shingles and the outer layer made from quality shingles or shakes.

Since we must replace the exact type and install these in the same fashion, we will need to make many notes where random ones are used. However, we probably will take several photos and have enlargements made. With these we can more accurately duplicate the original.

Finally, we must identify the finish or coloring of the shingle or shake. Again, in conversation with the owner, we can get answers to both questions. If the shingles or shakes were stained or prepainted at the manufacturer's, we should make an effort to obtain replacements from that source. If the shakes or shingles were stained by the builder or homeowner, we need to define the type and age of the stain.

In summary, we inspected the damaged shingle or shake siding and determined the type of shingle or shake, its original color, the technique used when installing it, the kind of substructure or surface, the method of trim around doors and windows, the cause of damage, and the pattern of installation. Our requirement is to renovate the walls that are damaged, restoring them to the original condition.

For our generic project, we will assume that damage was from a storm where flying debris ripped shingles free and then high winds got behind others and ripped them away as well. Damage was restricted to one side of the house whose surface measures about 270 sq. ft. The outside corners employ corner boards that were slightly damaged by the storm. The house has double-hung windows with pulleys and weight pocket. The tops of the windows have rain drip covered with copper that has many years remaining. The sheathing was not damaged, but the felt was torn and needs replacement. Our project has double-thickness shingles, even rows at 6 in. exposed. The undercourse of shingles is recessed 1/4 in. from the outer course.

Alternatives to the problem's solution. The alternatives to the problem for the owner are to have us replace the missing and damaged shingles by using an exact type match or we can remove the remaining shingles from the wall and reside the entire wall. A third possibility is to reuse the undamaged shingles and interweave them with new shingles, thereby decreasing the difference between this wall and the others. We need to discuss the cost factors with each solution. Generally, replacing just the damaged and missing shingles is least expensive in terms of materials and may be cheaper as to labor as well. However, the match will usually be different in color and, therefore, we will need to age them or stain them to match. This is not easy to do. If we replace the entire wall's siding, we first must remove the old shingles with no intention of saving them. This means minimum time for removal and no obstruction to residing the wall. Materials will cost more, but time efficiency will be best. If the owner agrees to have the old, reusable shingles used in residing the wall, then we need to explain about the added time removing the old shingles. But there would not be any added labor costs for the new installation. Also, the materials cost should be lower, which may offset the added labor costs.

Statement of work and the planning effort. Since the wall had damage where many of the shingles were ripped away and others were blown off, we must expect that some of those remaining are loosened and others are damaged. Our first statement of work must be to *test the remaining shingles for damage and sound anchorage.* Then we need to have the owner's consent to remove the entire wall if the damage exceeds 50%. Thus, *replace the entire wall covering and felt subsurface if the damage exceeds fifty percent of the square footage of the wall.* Another statement of work is to *select a colored shingle or apply color to shingles to match the present ones unless natural aging was the original coloring technique.* We also must *match the shingling pattern for style and character.* Finally, after discussion with the owner, we agree to *reuse as many serviceable old shingles as possible and interleave them with the new ones.*

Contract. The types of contract applicable to this job are the fixed price and fixed price for labor and variable price for materials. To offer a fixed price, we may require several days of inquiry to (1) locate the exact replacement shingle and (2) return to the site and determine if the damage exceeds 50%. We also need to remove a few old shingles to determine how many we can save versus how many we break. Then we can ascertain the quantity, price, and shipping costs, as well as a delivery date. The nails and felt needed and flashing, if required, are locally purchased. Since the statements of work clearly define the replacement technique and finish, we can and should repeat them in the contract. To this we add the labor costs for one carpenter who has the skills for shingling with wood shingles. We then add our cost of operation as to the job and finally a profit.

If we opt for the fixed price for labor and variable costs for materials, we should add the overhead and fixed and variable costs, to labor. Then we estimate the price of materials and shipping plus a fixed or specified markup. The owner may feel that he or she has no control over material purchases. So we can provide a materials price range or specify the percent of markup in writing. For this project, let's say that we estimate 40% of the old undamaged shingles can be removed without breaking. We do not plan to save any undercourse shingles. So we can estimate the costs of materials at three square of undercourse and three square of outer shingles less 40%. (There are 100 sq. ft per square.)

Finally, we must add the method of payment clause to the contract.

Material Assessment

Direct materials	Uses/purposes
Wood shingles	Replace the siding
Underlayer shingles	Undercoursing as required
Treated or aluminum nails	Nail felt and shingles
15-lb Felt (tar paper)	Subsurface behind the shingles
Copper flashing	Drip caps over doors and windows (if necessary)
Stain	Color or age the shingles

Indirect Materials	Uses/purposes
Frieze molding	May need to replace if broken while removing
Mineral spirits	Cleaner
Plastic sheeting and tape	Block windows from being sprayed

Support materials	Uses/purposes
Carpenter tools	Reconstruction
Ladder and saw horses	Scaffolding

Support materials	Uses/purposes
Planks	Walking surface
Paint spray outfit	Apply stain

Outside contractor support	Uses/purposes
None	

Activities Planning Chart

Activities	Time line (days)						
	1	2	3	4	5	6	7
1. Contract preparation	X						
2. Materials ordering and acquisition		X					
3. Siding removal and replacement			X	X			
4. Staining				X			
5. Clean up				X			

Reconstruction

Contractor support. This is a relatively simple reconstruction effort. Therefore, the contractor support activities at the office and on the job site should be minimal. Contract preparation and ordering supplies should be finished within a day or less. The first visit to the job site should not have taken over an hour or two, including travel. The second trip should be for the signing of contracts, and the third may be required to get the carpenter started. The final trip is for final inspection with the owner present. Then we can either present the bill for payment or, more likely, mail the bill in several days.

Materials and scheduling. The shingles should be relatively easy to purchase since shingles have been made by similar techniques for well over 100 years. The other materials are equally simple to obtain. The undercourse shingles would be ordered at two bundles per square and the top shingles would be ordered as grade 1, 2, or 3 at four bundles per square.

The carpenter would probably be required to pick up the materials on the way to the job on the day assigned to begin. Then he or she would complete the job in the expected 2 days and return on the next day to apply the stain to match the remainder of the house.

Shingle removal and replacement. The first part of our generic job is to inspect the shingles and determine how to proceed with removal of the shingles, with the goal of saving as many as possible. The reusable ones are to be interspersed with new ones. The technique for removing the shingles is to either pry the shingle away with a flat bar or use a nail set and drive the nail through the shingle. Then we lift it away and pull out the nails. It may be more difficult to save the undercourse since these are thinner and tend to split more easily; therefore, we make no attempt to save them. We should start at the cornice level and remove the frieze molding. If we are careful, we will not break any of the pieces.

After the wall is cleaned of shingles, we strip away the old felt and apply new felt. The felt must overlap the previous row in the direction of rainfall by 2 in., minimum. The side laps must be 6 in. or slightly more.

Then we begin at the foundation and reinstall the shingles. Since this is a double-thickness job, we first run a course of undercourse along the foundation line. Then we install the top layer, making sure to have a 1-in. offset at every joint. Since the opposite ends of the building are readily available, we can easily transfer the exact position of the bottom of the second and subsequent course onto the corner boards. We snap chalk lines across the wall, allowing for line sag, one course at a time.

The undercourse extends at least 2 in. into the third course. This is also true for the top ones. With a 6-in. exposed surface plus the second course and at least 2 in., we need to buy 16-in.-long shingles, but probably would buy 18 in. to get a quality grade. The nailing scheme would have a minimum of two nails per shingle. The nails must be about 1 in. above the bottom of the next course.

Cutting around the windows requires some top nailing under the sill. Treated or colored nails are used here to minimize their appearance.

The shingles should butt side to side, but they need not be jammed tightly. They will expand when it rains.

Staining. With the shingling job complete, we next make the shingles match the ones on the adjacent walls by aging or staining them. For this, we contact a paint supplier and obtain the correct aging stain or matching color. We apply at least two coats of stain, following the guidance provided from the supplier.

Concluding comments. In this project we examined the problem of restoring the cedar siding that was damaged on a house. We discussed several sheathing techniques we might find behind the siding, several different window and door installations, and the different techniques of applying shingles or shakes. Quality workmanship is the most important aspect of the job. With it, we can match the siding on the other walls and thereby complete a job fully satisfactory to the owner.

This concludes the chapter on restoration of siding. Many materials and techniques have been examined. The reader should be able to assimilate these descriptions to other very similar siding replacements, such as bonded aluminum.

CHAPTER SUMMARY

In this chapter we examined a wide variety of building materials that are the siding for residential homes. Each has its qualities and uniqueness of installation and maintenance. Some can be replaced by the homeowner who has a modest skill capability. Some require the expertise of the carpenter, siding installer or painter. Some of the problems are easily eliminated, while others such as replacing some siding pieces on a wall are much more difficult. We also discovered that the contractor spends considerable time obtaining materials that match the original materials. He or she also must supply scaffolding and tools not usually owned by the homeowner. Finally, the costs associated with restoration of the siding are usually moderately expensive. They must be used with as little waste as possible regardless of who installs them.

3

BRICK VENEER, MASONRY, AND STUCCO

OBJECTIVES

To clean face brick and make restoration to cracked mortar joints.
To eliminate the problem of a staggered crack in brick veneer or block wall.
To clean stucco and make repairs to decayed stucco.
To clean masonry units (blocks) and make restoration to cracked mortar joints.
To clean, reseal, and refurbish basement walls and exposed sections of foundations.

OPENING COMMENTS

If the siding is face brick, there could be several conditions that drive us to corrective actions. The most serious are the cracks in the wall resulting from settling. Other less serious conditions include decayed mortar joints and stained face brick. With regard to stucco, we have the possibility of serious problems, well beyond simple face cleaning. Due to the nature of stucco, water frequently gets behind the stucco and destroys it, as well as the sheathing and sometimes the structural members. Where block walls are damaged, the causes can be fundamental, such as cracked footings, or less severe, as in erosion caused by the effects of rain water. As with brick, cement and cinder blocks may have cracked mortar joints. But we also know that these units do not have the strength of brick and, therefore, they may split when subject to stress. The problems must be dealt with with the idea of eliminating the cause and making the wall sound.

As contractors, we will be responsible for examining the owner's home for damage to the walls when these are made from brick (or stone) or masonry units or when covered with stucco. We must be able to explain to the owner the cause for the damage, even when there is no apparent structural damage. Our discussions must be kept simple and straightforward, but detailed enough to convince the owner that we know what we are talking about. We may be either a masonry contractor or general contractor. The surest way of establishing our degree of expertise is to discuss the problem by showing the owner the extent of damage and to identify several causes that must be checked out.

The homeowner's expectations are usually twofold. First, he or she wants the damaged materials repaired quickly and cheaply. Second, the owner wants the restored wall to look like the original job. If the brick is stained, the stain must be removed. If the brick mortar joints are split, they must be repaired and the cause removed. If the wall has been stuccoed, but is cracked or stained or has fungus and general deterioration, we need to assure the owner that the repairs will make the wall like new. The owner who has either a basement wall that leaks or one that has cracks in it must be satisfied that the repairs will restore the wall to its original shape. The cause for the damage or discoloration must be removed to minimize recurrence.

In this chapter, we will study three projects: (1) problems and solutions with face brick, (2) problems and solutions dealing with stuccoed walls, and (3) problems and solutions dealing with block walls.

PROJECT 1. RESTORING THE FACE BRICK ON A RESIDENTIAL HOUSE

Subcategories include cleaning fungus from the brick; tuck-pointing the mortar joints; remortaring the loosened brick; tuck-pointing the face stone or Tennessee stone; eliminating the cause for cracked mortar joints above the doors, over windows, and near corners.

Primary Discussion with the Owner

Problem facing the owner. From the contractor's viewpoint, the cleaning effort is a minimal problem. He or she knows, from experience, that simple chemicals usually clean the fungus (black) growth from the brick or stone. The work involves the application of bleach or a product containing bleach with either brush or spray, followed by a clear-water spray or wash-down. Likewise, replacing eroded mortar in joints with tuck-pointing is also a simple task for an experienced mason. Not so easy is locating the reason for a series of broken joints (these are usually staggered down the wall). We know that there must be a stress point or two that has caused the cracks.

We know that the usual points where cracks occur are at the upper corner of window and door openings and several feet in from the corner of the wall. The severity of the cracks dictates the duration of the problem and increases the cost of resolving it. Cracks in mortar joints may be as much as 1/2 in. wide. We know that settling is one cause. Others include changes in footing heights, cracks in the footings, changes in foundations where vertical joints are made, and erosion causing breakage in the slab supporting the wall and veneer.

Alternatives to the problem's solution. There are only several alternatives to these kinds of problems. The owner can perform the cleaning him- or herself. The owner can ignore the cause of the cracked wall and have the mason make simple repairs, such as tuck-pointing all cracked joints.

On the other hand, the owner can require the contractor to eliminate the cause of the damage even if it takes reinforcing the footings or foundation. The owner can require the contractor to remove the brick where the cracks are and reinstall them with fresh mortar, after making corrections to prevent the effects of settling. These alternatives are more expensive, but attack the root cause and not the effects of the problems.

For our generic problem, we will treat a crack in the corner and over the door leading into the garage, as well as clean built-up fungus on the north wall.

Statement of work and the planning effort. After a careful study of the condition of the brick veneer siding on all the walls of the house, the owner and we arrive at several statements of work. The first is that *we agree to wash down the north facing wall and several lower sections of walls that are covered with fungus.* We further agree to *locate the cause for the settling and take appropriate measures to restore the footings to their original position and further take such measures to prevent further settling.* Next *we will either remove and replace the bricks affected by the settled area or, if we are able to move the wall sufficiently, remortar the cracked mortar joints.* Finally, *we will remove the several bricks over the cracks above the door and reinstall them using slightly wider mortar joints.* Our *work must employ quality materials and workmanship, and the finished job must conform to the remainder of the brick and mortar on the walls.* Finally, *no ground or surrounding areas will display any evidence that repairs have been made.* We translate this as ensuring a complete cleanup and restoration of gardens and shrubs, no residue of mortar mix, no evidence where trenching under the foundation was performed, and so on.

Contract. This set of conditions can make a fixed-price contract very difficult to formulate. By analysis, we can see that cleaning the surface of the bricks can be done with a fixed price once the approximate square footage is measured. This should be the first part of the multipart contract. We can check with the local office extension service to identify the type of fungus on the walls and with this information select the best cleaning agent. The method of applying the the solution and washing the surface will be by spray.

The problem with a settling or cracked foundation, footing, or slab is much more difficult to solve. We must first access the area, then determine the cause, and finally select a corrective action. Since there are no clear-cut answers, we will cost out the work on an hourly basis for the *problem definition* phase.

This means that we must program laborers to excavate around the foundation near the settled area in the wall. The hours will remain approximately the same for shrub and plant removal. But the work will differ significantly based on the type of foundation. If the structure is a slab and footing in one pour, then the digging is simple and not very deep. We will be able to determine the damaged area as soon as the footing is exposed. In regions where frost seldom occurs, the footing may be only 8 in. deep with a 4-in. slab on top. Figure 3-1 shows a cross section. There should be steel rebars in the footing, and the wire mesh should be fastened to the rebars. The bars should overlap according to building codes and be wired at the overlaps. Unless there is a quality inspection of these items, there is no way the owner can determine the quality of the slab and footing. If the crack is found, the damage will be evident.

The cause will not be the fact that the steel was or was not properly placed and wired or that there was sufficient steel. We know that properly prepared and poured footings and slab can carry some load. What we must look for is the reason the footing gave way. This is where the cause exists. The existence of damp ground or eroded ground under and around the foundation indicates high soil porosity and thus decreases the capacity of the earth to support the building. The combination of a weakened soil and poor quality controlled footing and slab construction resulted in the crack and settling.

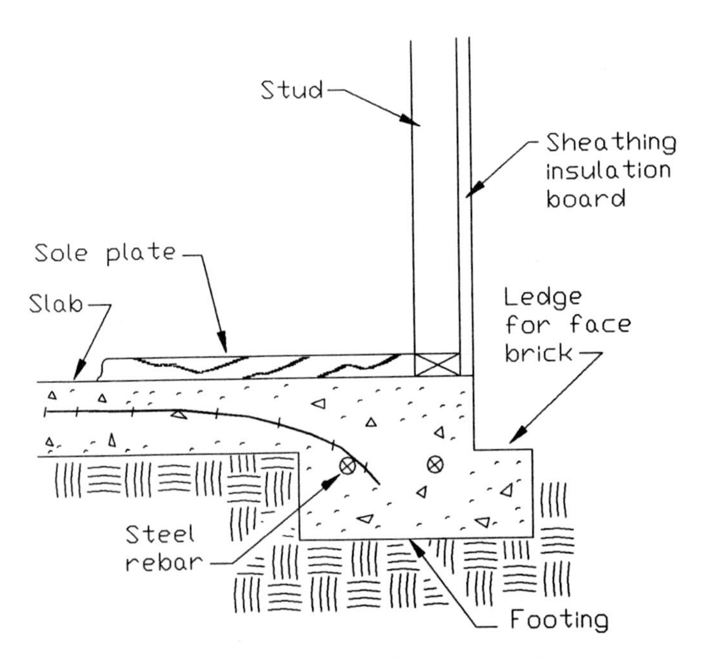

Figure 3-1 Cross section of slab and footing with rebars.

If the foundation is a crawl space or a basement with footings and block walls, the crack should travel entirely through the block and footings unless there was significant decay in the blocks supporting the brick walls. These walls will have a cross-sectional view as shown in Figure 3-2. Inspection of the wall will provide an indication of the magnitude of the problem and lead to its cause.

Fixing the problem can only be defined and estimated after preliminary work has been accomplished. Therefore, the second part of the contract would be best as time and materials, where time will be the most expensive aspect. In addition, we will probably use the help of a landscaper to remove shrubs and trees to make access to the work. This would be a cost to the owner, but it would be a subcontractor's work and cost to us.

A secondary contract between the owner and our company would be made after the exposure phase is complete. This contract may be either fixed price or a modified time and materials contract.

The third problem we contract for is the repair of the cracked mortar joints from the top of the door to the soldier course at the frieze. For this job, we will just clean out the joints and tuck-point new mortar in place. This third element of the contract is fixed price.

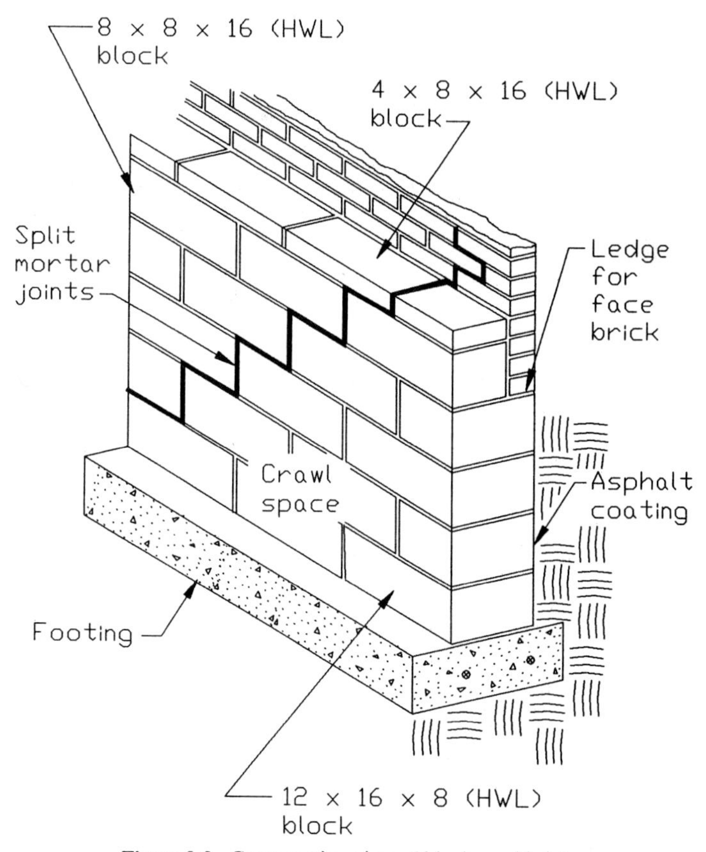

Figure 3-2 Cross-section view of blocks and bricks.

Let's summarize. The body of the contract will have three elements, each with a price and specifications.

1. Clean the bricks of built-up fungus, a fixed-price contract.

2. Perform exploratory work to locate the multiple causes of the long crack in the wall near the corner, a time and materials contract.

3. Repair the cracks in mortar joints between door head and frieze, a fixed price contract.

4. Subcontract for landscape service, a fixed-price contract.

5. Follow-on contract for repair to wall and eliminating the cause of settling, a time and materials contract with the possibility of limits to both.

Material Assessment

Direct materials	Uses/purposes
Chemical bleach	Clean bricks
Mortar mix and sand	Mortar for joints
Bricks	Replace broken bricks
Ground cloth or cover	Replace old ground cloth
Mulch	Soil preparation
Bark or ground cover	Replace the old cover

Indirect materials	Uses/purposes
Bailing cloth	Bound root systems of plants moved
Water	Wall wash and mortar

Support materials	Uses/purposes
Pressure-washer outfit	Spray the walls
Cement mixer or mortar box	Mix mortar
Masonry tools	Construction
Shovels, hoes, and rakes	Excavation and mixing mortar
Scaffolds and planks	Work platforms

Outside contractor support	Uses/purposes
Landscaper	Move and reestablish gardens

Activities Planning Chart

Activities	Time line (days)						
	1	2	3	4	5	6	7
1. Contract preparation	×						
2. Exploratory work		×	×				
3. Repair of cracked wall			×	×	×		

(*continued*)

Activities	1	2	3	4	5	6	7
4. Repair of cracked courses					×		
5. Removal of fungus						×	
6. Remove and reestablish gardens		×					×

Time line (days)

Reconstruction

Contractor support and scheduling. Estimating this contract and preparing it with the variables stated above will require some planning since various types of personnel are expected to work on the job. For example, we can program the laborers to perform the excavation and assist the masons with efforts to locate and correct the cause of the settling. They will also assist with the reconstruction effort by preparing the mortar and maybe removing the brick. In addition, they can also spray the walls to remove the fungus.

The lead mason should be on the job during the exploratory work to guide the laborers and remain on the job throughout the duration. Other masons would be required for those days when bricks are installed and mortar joints are tuck-pointed.

The subcontractor must be on the job site the first day to carefully remove the shrubs in the gardens. Then he or she must return to reestablish the gardens when all the repairs are complete.

The office must prepare the contracts and order the materials. The contract for the owner will have to wait on the subcontractor's contract bid. Then the contracts must be signed by the owner. Company overhead must be included as well as markup for profit.

Someone in the office must also schedule the job and the workers and ensure that the materials are picked up a day ahead of need or are programmed from company supplies. The subcontractor must also be informed of the start date in sufficient time to permit his or her scheduled activities.

Exploratory, repair of the wall, and subcontractor work. We have already discussed the essentials of the work to be done in this phase of the job. Our primary goal is to locate the cause of the settling, which caused the crack in the brick wall. We only need to expose the foundation and footing to determine if either is cracked, and we look for washed-out, poorly packed subsoil or other reasons for the damage. If the result of the exploration points to subsoil that is not fully compacted up to and solidly in touch with the footing, we suspect that the weight of the roof and wall was too heavy for the foundation and footing to sustain. This can happen when the house foundation and adjacent garage foundations are built as separate units and at different heights, as shown in Figure 3-3. The bricks are laid up so that the rows are even. But the garage settles and pulls away from the house. A cracked wall occurs. Let's assume this is the situation. What can we do about it? Depending on the severity of the damage, we can ensure that further settling does not occur by adding reinforce-

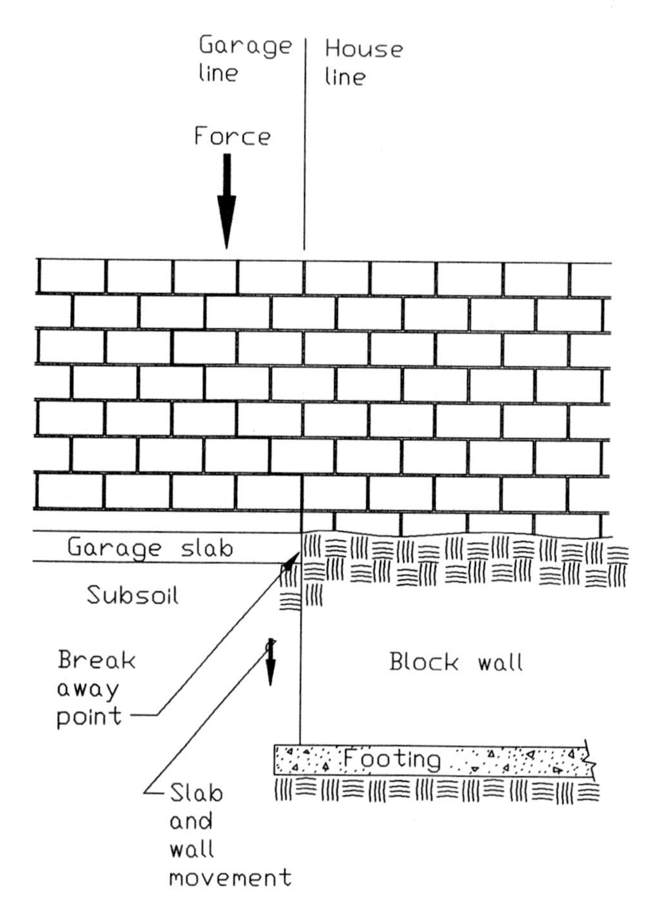

Figure 3-3 Brick walls with foundation and slab; crack at joint.

ment under the slab or foundation of the garage. It will not correct the separated brickwork, but that is the next phase of the job.

We can attempt to raise the wall and then pour pilasters to support it. For the raising job, we may need the assistance of house movers, or we would need shoring and house jacks. The idea is to burrow under the footing and install heavy shoring timbers onto solid ground. Then, with the aid of 10-ton house jacks, we slowly raise the footing until the courses of brick align horizontally. Once in position, we form the area for pilasters and fill them with concrete. After the concrete sets and cures (about 2 weeks to 1 month, we will use steel shims between the pilasters and slab or footing to fill the remaining space if required.

If the settling was slight and less than 1/2 in., and we are convinced that all settling has occurred, we can advise the owner that the best solution is to remove the bricks from the wall and reinstall new bricks. While we do this, the masons graduate each row slightly to make up for the previous condition, or they make up the difference below the first course.

Repair of the cracked brick wall covering. Repair of the brick walls that have split mortar joints or split bricks along with split mortar joints is the job we are faced with. We must evaluate the degree of damage and select one of the following solutions to make repairs.

1. Cut out the old mortar and tuck-point new mortar if the rows are in alignment and the separation of the joint is very minor.

2. Remove several bricks from either side of the crack with the idea of reinstalling them with slightly wider vertical joints to make up for the wide split.

3. Remove all rows up the wall where mortar and bricks are split vertically and cracks are also in mortar joints horizontally; then reinstall them, making adjustments to spaces as needed and replacing broken bricks.

The mortar we use must be appropriate for the job. On almost every job, we use the standard mortar mix along with sand and water. But, in some situations, we would use quick-setting mortar or add coloring agents if called for. The mason laborer will prepare the mortar to the correct consistency.

The skilled mason will lay up the brick to overcome the problems. For example, in our scenario where the footing slab was raised until the courses were level, the mason could perform either one, two, or three courses above according to the severity of the cracks in the brickwork. In our earlier description, where the bricks above the doors to soldier course separated, he or she would probably use solution 2 or 3 to make the repairs.

Removal of fungus. Although there could be several causes of discoloration on the brick, we opted for the treatment of fungus. But, after discussing how to treat for fungus, we will discuss how to remove stain from other sources.

The most common treatment for fungus is to apply a full-strength solution of bleach to the wall. This can be done with a high-pressure water unit if the contamination is extensive or with a brush or broom if the area is small. Laundry detergent mixed with the bleach will dislodge the dirt and adds a small amount of foam. The foam makes it easy to see where the wall has been treated. Commercial products are also available at various building supply outlets and paint stores. Some are in concentration form and are mixed with water before application.

After the application has set for several hours, the brick wall should be washed with water. If the fungus is not entirely removed, we repeat the process.

Reddish-brown stains indicate iron, which could come from rusting rebars, metal L-shaped lentils across windows, or rock that contains iron oxide. Chemicals easily remove the stain. We would then apply some form of sealer to prevent air and moisture from mixing with the iron to form rust. This could be a clear silicon spray, mortar, or caulking.

Reestablish the gardens. The gardens have been disrupted as a result of the exploration digging, cementing the joints, reinstalling the bricks, and other work. The ground has been contaminated with chemicals from cement, subsoils, building materials, and chemicals from washing the house down. The gardener must reestablish the soil first and then replant the shrubs and apply the ground cover. The gardener will know what to do, but let's make some judgments about his or her activities.

One thing that must be done is to condition the soil according to the type of plants and shrubs to be planted or replanted. If they are acid loving, such as evergreens, azaleas, hawthorn and others, then pine mulch and acidic enhancers would be added to ensure a proper medium for growth. If the plants require a more alkaline soil, then the Ph would need to be between 7.0 to 7.4.

Besides adding chemicals, the gardener would add mulch such as peat moss, pine mulch, cypress mulch, or a combination of these to cause the soil to retain moisture and have a certain amount of porosity.

When the gardener replants the shrubs, they will need to be planted to their previous depth, and the soil must be saturated with water to remove all air surrounding the root ball. This elimination of air is very important for the resetting of the plant. Also, a technique used in planting shrubs is to build up a soil ring around the plant at about the perimeter of the branches. This soil buildup acts like a dam and permits water to be trapped and retained for the plant. When the plant is reset, more mulch is usually placed around the plant. However, since the garden is being rebuilt, the gardener will likely apply mulch after the shrubs are all set in place and the ground cover is laid. This makes good sense.

Concluding comments. We have examined several important principles concerning the care and maintenance of brick walls. A settled wall will surely crack some of the joints. In extreme cases, the bricks will crack as well. The obvious points of stress are where two different foundations join and over the lentils of doors and windows. Considerable pressures are applied to these points from the weight of the walls and roof. We also examined the cause of damage from subsoil erosion.

We determined the type of contracts that would be appropriate and fair for the contractors and owner. We pointed out that where unseen situations exist we must amend the contract when more information is available. This is after exploratory work is completed.

PROJECT 2. REFURBISHING STUCCO

Subcategories include replastering with cement plaster; applying wire reinforcement or other types; premixed stucco treatments; replacing damaged wood under the stucco; repairing cracks in stucco; installing expansion joints; painting stucco.

Primary Discussion with the Owner

Problem facing the owner. When a homeowner calls the contractor to examine his or her stucco-covered home, there could be serious problems that are very expensive to correct. When stucco is first applied to the exterior walls of a house, certain conditions are usually met that ensured long, trouble-free service. Figure 3-4 shows a cross section of a stuccoed wall that has wood framing and wood sheathing.

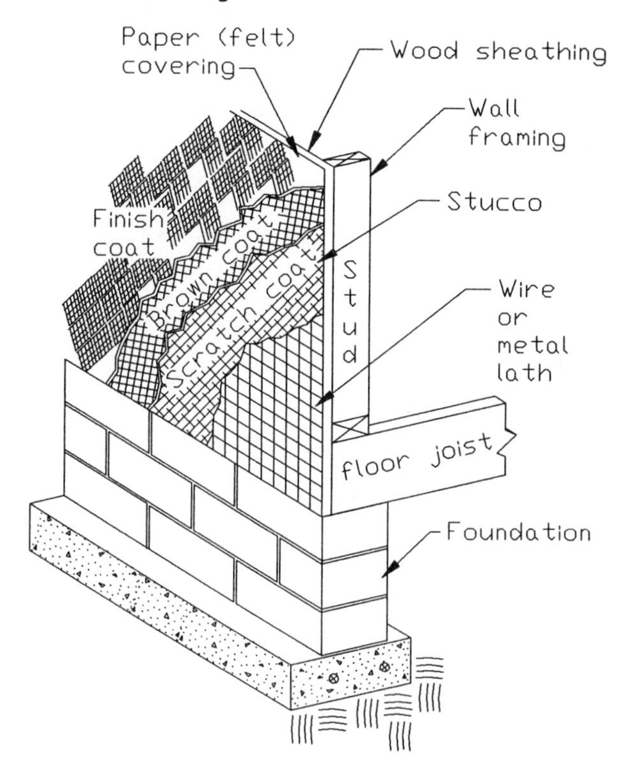

Figure 3-4 Cross section of stucco wall; wood framed.

This is the one we shall be discussing. However, we need to illustrate the difference when stucco is applied to block walls. Figure 3-5 shows the details. (Take time to study the details provided in each figure and note the names of the materials and different coatings of stucco.)

The fact that we are called to examine the problems usually means they are well advanced. Some obvious conditions that we should expect to see include the following:

1. Cracks in the surface, which may include contaminants.

2. Discolored finish, which can be a clear indicator of water damage from the base coat absorbing water through leaks.

3. Bulges in the wall, which can indicate that the reinforcement materials have deteriorated and have broken free from the studs or sheathing or that the mechanical key or suction has failed.

Hairline cracks and wider ones may result from movement of the house caused by ground shifts or high winds such as tornadoes and hurricanes. The movement stresses the cementitious materials beyond their capacity to expand. If the house is newly built, the cracks are more than likely caused by improper curing techniques at the time of application. Most mason contractors ensure that expansion joints are

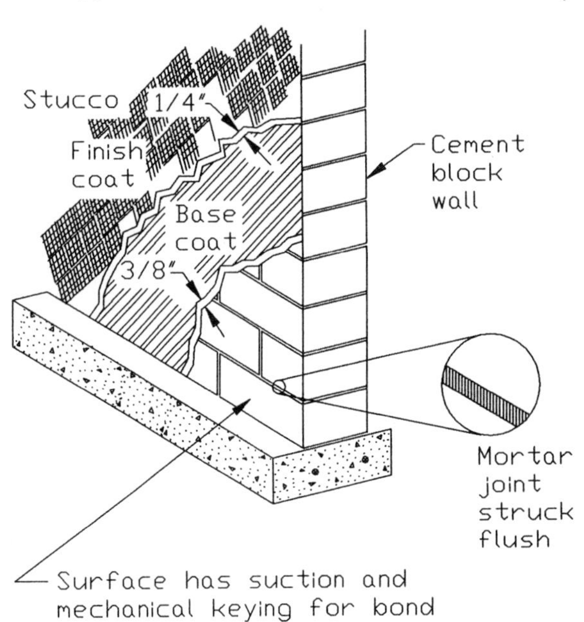

Figure 3-5 Cross section of stucco wall; block masonry wall.

strategically placed to meet recommendations where a segment does not exceed 150 sq. ft. For example, if the wall height is 10 ft, then expansion joints would be placed not farther apart than 15 ft.

As we examine each separate condition and take notes, we discuss the construction details and probable causes with the owner. Hairline cracks without contaminants can usually be covered with a fresh coat of appropriate paint. The more serious cracks are those that have existed for some time and are contaminated with organic matter. They leak and so water has entered the brown coat and maybe even into the scratch coat. These are the two base coats of the three-coat system. The top or finish coat is different and contains one of several types of lime in a cement and sand mixture. When water enters the brown or scratch coat, it deteriorates and its bonding capacity decreases or fails. When the condition progresses sufficiently, pockets of loosened stucco develop. These are found by looking for bulges or by tapping the wall and listening for hollow sounds.

If the cracks in the stucco are old, we need to probe to determine their depth and the extent of further damage if any. The owner must understand that this examination is essential. Since the materials are basically cementitious, consisting of portland cement, masonry cement, plastic cement, and lime, they are very hard, and we need to use a steel chisel to chip away along the cracks. We can tell if water has entered the cracks by the color of the base coats, because the color will be darker than noncontaminated base materials. We may be able to fill the cracks with finish coat, or the entire wall can be refinished, which will fill the cracks as well.

Where the wall feels spongy or moves in and out or where there is clear evidence that stain has bled through from the back, we have the most serious problem and the most costly as well.

Alternatives to the problem's solution. We must now examine the alternatives available to the owner for each of the conditions we might have encountered. For minor cracks devoid of contaminants, we can offer a cleaning of the surface with acid and application of a new top dressing or the application of an exterior paint system.

Where fungus exists and contaminants have lodged themselves, in cracks, we must remove these with tools or solvents or both until none exists. Then we need to either apply a top coat or, first, a brown coat followed by a top coat in the damaged areas. The owner needs to know that this treatment will require a paint system to match the other parts of the house or a colored finish coat applied over the entire area.

For these extensive cracks, the alternative is to cut an expansion joint into the wall in one or more places to permit expansion and contraction of the new applications. We would advise the owner of the places and number of joints and attempt to show in a sketch how the new wall will appear. If we are careful and place the joints below and above windows and doors, they will be more pleasing than in the middle of a clear-wall expanse. We can also apply proper curing controls for resurfacing the wall to eliminate the stress normally caused by shrinking during curing.

If the damage to the stucco extends to the base of the wall and the metal reinforcement is damaged as well, we have no alternative but to remove the entire damaged area and begin over. In this situation, we can recommend the installation of control joints to limit the amount of wall to replace. This does two things: (1) it eliminates a patch job, which may not exactly match the rest of the wall, and (2) it permits us to remove more of the wall area to make repairs simpler and ensure that the cause, as well as the effects, is removed.

Statement of work and the planning effort. In our generic project, we are going to make restoration to several parts of the walls. Each has a different and more serious problem. The first statement of work is *to wire brush all cracks and identify those that must be cleaned and scored for repair.* The next requirement is to *remove all sections that have evidence of moisture contamination, to and including decayed wire, and replace the materials.* If creating expansion joints is part of the solution, we would include these as well. The *final finished products must match the remainder of the house for color and texture even if the entire house must be sprayed with paint after repairs are completed.* Furthermore, *while repairs are being made, all necessary precautions must be taken to prevent further damage to other parts of the house and grounds beyond the work areas.* If there is damage to the sheathing, subflooring, or sills, for example, *All damaged wood framing and other materials will be replaced as required, if evidence of damage is found.*

The planning effort for this work takes the combined efforts of the estimator, who will make the actual measurements and determine the quantities of materials needed to effect repairs and the personnel needed to complete all the work, and the contractor to prepare the contracts. The weather will also be an important consideration because the various layers of stucco will require protection from rain and cold weather. Another planning factor is the total number of worker-days needed to replace a section of wall. Proper setting and curing times must be allowed between coats of stucco. This may translate into one or more days with partial idle time. Workers expect to be paid a full day's wage regardless of the actual time used. The types of personnel we will need for the restoration include masons, mason laborers, and very possibly carpenters. Stucco materials will be mixed on site, so the time for picking up and transporting them to the site will be figured into the time line.

Contract. Based on our definition of the types of problems to solve and their extent, plus examination of the conditions found at the site, we can prepare a fairly accurately fixed-priced contract. With one exception, which is replacement of wall sheathing and damaged framing, we can offer a fixed-price contract that includes allowance for overhead, rental equipment, profit, and subcontractor support. The owner would receive a contract bid detailing the reconstruction on each wall.

We prepare the standard elements of the contract to include the parties and location of the job and a clear description of the statements of work that we identified earlier. The one clause that must be included concerns unknown factors about damaged or rotten wood products, and it might look like this:

> **Exception clause:** Replacement of damaged sheathing and framing materials will be priced separately. The price out will be agreed to after a clear assessment can be made with the owner present. This will include labor and materials. No alteration to the original contract price will be effected unless the repairs require exposing more of the wall than originally specified. In that event, the basic contract price will be proportionally increased to cover the added work and materials.

Material Assessment

Direct materials	Uses/purposes
Metal lath	Expanded or wire lath used as reinforcement for stucco
Control joint	Preformed metal expansion joint
Corner reinforcement	Preformed metal external corner material
Flashing	Over windows and doors and over rain cap above foundation
Backing paper	Felt or building paper that separates the stucco from the sheathing
Type I, II, or III cement	Part of the stucco mixtures
Masonry cement	Suitable for stucco
Plastic cement	When used, eliminates the need for lime
Sand	Part of the stucco mixture
Water	Needed to mix the stucco
Quicklime	When added to cement, forms a lime–cement stucco
Nails	1½-in. Roofing nails
Sheathing	Replace the damaged sheathing

Indirect materials	Uses/purposes
Plastic sheeting	Protect exposed surfaces, scratch coat, brown coat from rain and cold
Mortar board	Laborer places scratch or brown coat on this for ease of mason's use
Sandblasting sand	Cleans masonry walls of contaminants prior to applying new stucco

Support materials	Uses/purposes
Scaffolding	Simplifies work
Sawhorses and planks	Eases work for masons
Hoses, mortar box	Mix stucco and clean tools
Mason tools	Construction
Carpentry tools	Make repairs
Stucco scratch	Make scratch marks in preparation for the brown coat
Wide brush	To dash apply a bond coat if required

Outside contractor support	Uses/purposes
Landscaper	Remove and replace the landscaping to permit masons and carpenters access to walls
Carpenter	To make structural repairs prior to applying new stucco

Activities Planning Chart

Activities	Time line (days)								
	1	2	3	4	5	6	7	8	9
1. Contract preparation	X								
2. Scheduling and materials	X								
3. Removal of damaged walls		X	X						
4. Replacing wood materials			X	X	X				
5. Installing metal products				X	X				
6. Applying scratch coat					X	X			
7. Applying brown coat						X	X		
8. Applying finish coat							X	X	
9. Landscaping		X							X

Reconstruction

Contractor support. From the activity list, we see that contractor support begins with contract preparation, scheduling, and obtaining the materials for the job. The contracts are prepared for the owner and subcontractors. The owner commits to the job by his or her signature. The subcontractors must not only bid for the job but be scheduled according to the times they are required. We see that the carpenters are needed during days 3, 4, and 5. The landscaper must remove the scrubs on the second day, the first actual day of reconstruction, and replace them on the last day.

The materials must be ordered and picked up from the building supply company. Due to the volumes of cement and sand necessary, the supplier would probably deliver them to the sight. According to the activity list, this means that the lumber, cement, and sand should be delivered on the afternoon of day 2. This will preclude lost labor.

Removing damaged materials. The carpenters will remove the damaged sections of stucco by first making vertical cuts with a saw. They will remove all the sheathing to expose the stud wall and floor framing to assess the extent of repairs. At this time, the owner and contractor must agree to the conditions of the contract and arrive at a price for the carpenter work. The carpenter contractor will be the one who has the most direct input to time and materials, plus his or her allowance for overhead and profit.

All damaged wood materials must be removed and replaced, but care must be used to avoid damage to adjacent materials and interior wall coverings.

Paper-backed reinforcement and wire and accessories. After the old damaged stucco has been removed and in those places where the carpenters replaced wood materials, the masons will first install backing paper. This paper must be waterproof. Either roofing felt or building paper meet this requirement, yet both are vapor permeable. All pieces must be installed so that overlaps are in the direction of rainfall. Laps at corners should extend 6 in. around the corner.

Next we install the metal reinforcement. Several types and styles are available and include the following:

1. Expanded metal lath, which is available in diamond mesh, stucco mesh, flat rib, and 3/8-in. rib lath.

2. Wire lath, which is available in woven wire and welded wire.

Several of these are available in plain, self-furring, and paper-backed types.

Once the wire is installed, we will install corrosion-resistant flashing where required. This might be made from copper or galvanized or stainless steel metals; a newer selection would be plastics.

We will also install the corner strips or beads and any other control joint materials at this time, as well as drip screeds and casing beads if required.

Applying stucco. Steps 6, 7, and 8 in the activity list identify the three-step application process. This process is always used where wood framing is the foundation.

(In the case of stucco on masonry materials, a two-step process is used. See the next section for more on the two-step process.) The masons will mix a batch of scratch coat of stucco, which is largely a mixture of portland cement, sand, and fiber. It is applied with trowel, and before it dries the mason will use a piece of scrap wire as a scratch. With this tool, he or she makes deep scratches into the fresh first coat. Hence the name scratch coat. The scratches provide a better bonding for the next or brown coat.

The brown coat is applied over the scratch coat, but first the mason will lightly wet the scratch coat to improve bonding. As with the scratch coat, the brown coat is applied 3/8 in. thick. Certain skills must be brought into play when applying this coat. For example, entire sections should be done at one time without stopping. The application also includes screeding or rodding the surface and then darbing it to make it as level and uniform as possible. When complete, there should be no deep scratches and no uneven textured surfaces, which would prevent a finely applied final coat.

The final or top coat is applied last at a thickness of 1/8 in. It can be site prepared or purchased premixed in 5-gal or larger cans. This finished coat can be textured to the owner's specifications. However, in our contract we are obligated to match the texture of the remainder of the house.

Two-step stucco process. In the two-step stucco process used on foundations of masonry, the base coat needs to be 3/8 in. thick and is applied directly to the masonry. The masonry needs to be slightly moistened to ensure proper bonding. In addition, the surface of the first coat must be uniform and level and have about the same texture. To apply the finish coat, we first dampen the base coat and then apply a 1/4-in.-thick coat from either premixed and on-site mixed materials (see Figure 3-5).

Landscaping. Landscaping the property is the final step in this contract. Recall that the landscaper removed established plants for replanting after reconstruction is completed. The first important step for the contractor is to clean the soil of all masonry debris. Cement, stucco, and its residue contaminate the soil and change its Ph. The contractor should also perform a soil analysis to ensure a proper soil balance for the types of shrubs used. Generally, chemicals, peat, and fillers are added to enrich and restore the soil's composition.

Then the contractor will set the plants in place and apply the contract-approved ground cover.

Concluding comments. We have examined the problems that could face the contractor when a stucco-covered house is found to be in disrepair. In our generic problem, we created the worst of possible situations. The reader can use this problem study to establish an approach to satisfy the owner's problem. The complexity of designing a contract that has both fixed-cost and time and materials sections may add to the overall difficulty of the job. Besides the landscaper as subcontractor, the carpenters may also be subcontractors. Finally, weather is a big factor in this type of work. Therefore, selecting the best part of the year or month may be important to the contractor's success and homeowner's satisfaction.

PROJECT 3. RESTORING MASONRY WALL INTEGRITY

Subcategories include repairing cracks in mortar joints; replacing decayed blocks; rewaterproofing below the ground surfaces.

Primary Discussion with the Owner

Problem facing the owner. As a rule, the homeowner will tolerate damaged block walls for a long time before resorting to the help of a contractor. In many situations, he or she will make all kinds of patches to stop cracks or to stop up the cracks that are leaking. In dry years there will be little further damage, but in rainy or wet years the problems will get worse. The problems facing the owner may be found in basement walls as well as those above the ground level.

When, as a contractor, we are called to the homeowner's property, the problem is usually well advanced and a solution must be developed. Nothing the homeowner has done will either fix the problems or prevent further damage.

We might find several conditions and we will need to examine each and discuss them with the owner. As always, we must be honest and provide sufficient information to the homeowner so that he or she is confident that we are telling the whole story.

The first of these conditions is a crack in the wall that runs several courses. This may be vertical or horizontal; but more than likely it is a combination of both. If the joints are cracked but the blocks are not, the owner is fortunate, since we can usually remortar the joints and strike them. However, if the blocks are split as well, we have a serious problem that can only be corrected by removing and replacing them. This is much more expensive.

Another situation we may encounter is a wall that is leaking water. Blocks are very porous and will permit groundwater to penetrate. Concrete blocks retard water better than cinder blocks, but both require the application of a bituminous coating to prevent seepage. If the wall is leaking water, we must advise the owner that something has failed, and it may not be the blocks. There is a strong likelihood that the coating used to prevent water penetration has failed. The solution could be as simple, but not cheap, as removing the earth from the wall and, after cleaning the wall, applying one or more coats of bituminous. Then we would backfill and replace the shrubs. Although simple, the actual trench might need to be 2 ft wide and 6 ft deep to permit a worker room to clean the wall and apply the new coating and to reach down to the footing. We will also add gravel to make a path for the water to run off before backfilling the trench. This job is labor intensive.

If we encounter crumbling blocks from whatever cause, we are obligated to inform the owner that these must be removed and replaced. We should attempt to define the cause of the decay, although it was most probably from water penetration, which later froze and thus ruptured the bond of cement and sand within the block.

In our generic problem, we will treat several of these problems.

Alternatives to the problem's solution. We identified three likely causes of problems with block walls. The alternatives available to the contractor in solving the problem include the following:

If the cracks in the wall are unsightly, but have remained unchanged for several years, the probable cause was stress at the time of construction or shortly thereafter from settling. We could offer a simple solution of applying either a two-step plaster job similar to stucco or a thin coat of cement plaster modified with a plastic bonding agent. This would eliminate the unsightly cracks and provide a smooth wall suitable for painting.

If the walls leak, unfortunately, we must attack the problem from the outside rather than apply plaster to the inside. However, instead of reapplying bituminous material, we could apply an epoxy or plastic material with a spray gun after the wall has been cleaned. We could also attempt an inside solution: we could apply a plastic sheet glued to the wall with an adhesive that is similar to epoxy. The wall would still absorb water through the block; we would need to check for water leaks at the point where the wall meets the floor.

The alternative to replacing the crumbling block could be to apply several coats of cement plaster with plastic bonding agent. Each coat would be about 3/8 to 1/2 in. thick. This solution is used only if the repaired surface does not have to meet the character of the wall, for example, if the repair is done to blocks in the back of a closet in the basement or to the exposed blocks above ground level where a plaster coat would normally be applied.

Statement of work and the planning effort. For our generic project, we will assume that water penetrates the wall below the exterior ground level and there is a crack in the wall that has a separation of 1/4 in. above the exterior door. The first statement of work is therefore *eliminate the cause of the water leak through the wall and make repairs to the wall surfaces as required.* The second statement of work is *tuck-point the joints where the mortar joint has separated and remove and replace the cracked cement blocks.* Since there will be discoloration around the repairs to the blocks above the door, we must apply a painting system to both the outside and inside of the wall. The statement of work is *after making repairs to the walls, apply a matching paint appropriate for cement walls.*

The planning effort is much like the one where brick walls had problems requiring repair. Here, we must employ laborers to excavate the ground from the wall to uncover the cause of the seepage. We must also use skilled masons to make repairs and painters to apply the painting system. We might have to use a landscaper to remove well-established shrubs before the digging begins and then restore the gardens later. In summary, three separate skills are involved, and we might have to use separate subcontractors for each effort. The various kinds of materials for this job are few, so their costs should not only be minimum but the time to acquire them should also be short. One main concern is the timing: we will require several days of dry weather. This timing effort will require the coordination of the subcontractors to a limited extent, since no one part of the work effort causes problems that could further increase damages to the present structure.

Contract. As indicated previously, we must prepare a single contract for the owner. Then we must have separate contracts for the masons, painters, and landscapers. Let's prepare the body of each as a bid/contract type with a fixed price.

General Contractors Contract. The body of the contract will state:

> For the fixed price stated below, we will perform the restoration to the walls affected by leakage and cracks. The water problem causing the leakage will be eliminated and provisions for removal of the cause and sealing of the wall below the surface will be performed. With regard to the cracked blocks and mortar joints over the door, we will restore the materials to weathertight and appearance. Further, the interior and exterior walls that require work will be painted with a painting system appropriate for cement. All landscape and gardens disturbed will be restored.

Mason's Contract. The mason's contract will be the most costly of the subcontractors, since they will perform the work of excavation and all masonry work. The body of the contract will state:

> For the fixed price stated herein, we will eliminate the leakage problem associated with the wall that is partially underground. We will seal the exterior surface again. We will remove cracked blocks over the door and replace with new similar ones. Other cracked mortar joints will be resealed appropriately. All residue of masonry materials will be removed from the job site.

Painter's Contract. The painter will have the next most costly subcontract. The body of the contract will state:

> All interior cement block walls shall be prepared for repainting after the the masons have completed their work. The interior walls shall have one base coat and two top coats. The entire area of exterior walls, including those that have been repaired, will be primed and painted with an appropriate exterior latex system. All supplies and residue of materials will be removed from the job site upon completion.

Landscaper's Contract. The body of this contract will state:

> For the fixed price indicated herein, we will remove the established shrubs near the masonry wall to be repaired and preserve them for later replanting. After the masonry work is done, we will reset the shrubs and relandscape the surface by applying appropriate nutrients and ground cover.

Material Assessment

Direct materials	Uses/purposes
Cement blocks	Replace the broken ones
Mortar mix and sand	To mortar the blocks in place

Direct materials	Uses/purposes
Bituminous coating	Seal the wall after cleaning
Fill dirt	Refill the excavation
Cement wall painting system	Interior and exterior systems
Cement plaster sealer	Interior wall sealer
Mulch, peat moss, top soil	Gardens
Fertilizer	Gardens
Bark	Ground cover

Indirect materials	Uses/purposes
Plastic sheet	Protect floors from mortar and paint
Bituminous brush	Apply asphalt to block walls
Burlap	Wrap shrub balls
Roller and pads	Expendable roller and pads for applying paint
Masonry blades	For power saw

Support materials	Uses/purposes
Mortar mixing box	Mix mortar for masons
Mortar board	Mason's surface for mortar
Scaffold and planks	One-level scaffold and minimum three planks
Shovels, hoes, hoses, rakes	Digging, mixing mortar, etc.
Power saw for cutting blocks	Includes blades
Power paint applicator	Optional painting equipment

Outside contractor support	Uses/purposes
Masons	Masonry work
Painters	Repainting work
Landscapers	Restore gardens

Activities Planning Chart

Activities	Time line (days)								
	1	2	3	4	5	6	7	8	9
1. Contract preparation	X								
2. Scheduling and materials	X								
3. Gardening		X							X
4. Seepage and sealing problem			X	X					
5. Cracked wall problem					X	X			
6. Painting system							X	X	X

Construction

Contractor support. The general contractor must provide the bids for the subcontractors. Therefore, the estimator or contractor must prepare the statement of work in sufficient detail for the subcontractors to adequately submit fair contract prices. The contractor may also require each subcontractor to make an on-sight inspection of the work to be performed and submit an estimate. This second approach carries the risk of greatly misunderstanding the work agreed to by the owner and general contractor.

The contractor will probably have the subcontractors meet at the job site and explain and discuss the work expected from each. This ensures a clear understanding and makes initial scheduling and alternate scheduling more reliable for all parties.

The contractor needs to plan to be on the site as each subcontractor is performing to assist as necessary. He or she need not spend much time there on any given day. The contractor must also be on the job at the conclusion of each subcontractor's effort to ensure quality work and proper cleanup.

The contractor must also allocate office expenses to cover the time associated with the job and all the billing and subcontractor contract preparation and filing. All other variable costs must be allocated as a factor of the total cost or by percentage of labor.

Materials and scheduling. The general contractor will not need to support the subcontractors with materials in this job. Each will be bidding for their own part of the total effort. Therefore, the contractor only has to provide a schedule and alternate schedule. The time line indicated above is very likely the one most suitable. The job can be completed in 9 days if everything works right. But delays are possible due to unforeseen problems with the water problem and weather and conflicts with contracts that the subcontractors already have.

The schedule shows the average number of days each subcontractor should stay on the job. Notice that, except for the landscaper, each can stay with the job to completion if weather permits.

Gardening. Briefly stated, the gardener must arrive on sight before the masons to remove the shrubs from the area that is to be trenched. All well-established shrubs must be protected and watered to prevent shock and damage.

After the work by the other subcontractors is complete, the gardener must return and reestablish the beds for the shrubs. He or she must appropriately fertilize the soil after ensuring that the base is rebuilt with peat, mulch, acidic additives if required, and additional top soil.

An appropriate ground cover must be applied. This must be the same as used on other beds around the property.

Seepage problem. For this part of the mason's job, the laborers will remove the earth next to the foundation to create a trench about 2 ft wide and down to the footings. The purpose is to reveal the cause of the seepage into the basement. It is quite likely that an underground stream has formed, or it may simply be that the groundwater level is high enough to place water against the wall. Our solution is simple. We must provide a drainage made of gravel and, if necessary, field drain 4-in. plastic pipe to allow the water to drain away from the house.

We must also scrub down the wall to remove all dirt. For this job we may also require the use of a high-pressure steam or water spray to dislodge the sand, dirt, and microorganisms.

Next we will apply several coats of bituminous asphalt with brush or mop to seal the block wall. Before applying the coating, we carefully inspect the blocks for decay, crumbling, and cracks. We also inspect for loose or damaged mortar joints. If these conditions are found, we will require the owner's consent to modify the contract on-site. When approval has been secured, we will correct the masonry problems and then apply the coating.

Split joints and blocks problem. The second problem the masons must deal with is the area of expansion above the door leading into the basement. This stress crack has become unsightly and a cause of leakage that stained the interior. As Figure 3-6 shows, the customary construction of block wall should have been sufficient to prevent this problem. But that was not the case. The top course of blocks is a solid cap course. The lintel blocks over the doorway have been poured full; the crack is just outside the lintel.

For the corrective action, the masons must remove the broken blocks with a chisel. To make a work surface, the laborers will erect scaffolds, set up the mortar mixer, and bring the sand and mortar mix to the job in readiness for the masons. After the blocks and old mortar have been removed, the masons will install new ones and, at the same time, they will tuck-point the cracked but cleaned mortar joints.

Since this work requires access to the basement to complete and strike the joints, the floor of the basement must be protected with plastic sheet goods. We will also protect any sidewalks from mortar while working outside.

Painting system. The painter has several options to restore the walls. We will examine these. The choice of finishes ranges from flat to satin to semigloss. The paint vehicle best suited to cement and cinder block is acrylic latex.

Since we have two situations, one involving wetness from seepage and the other new blocks and fresh mortar, the painter must wait at least 30 days for drying and curing before the paint can be applied. Then block filler should be applied first to the new blocks and mortar joints that were repaired. This primer is heavy and only one coat need be applied.

When painting the previously painted walls, inside and outside, the painter must follow a specific routine of first cleaning the walls, then testing the wall's paint adhesion, and finally applying a masonry conditioner coat followed by two top or finish coats. To clean the surfaces from smoke, oil, stains, mildew, efflorescence, old paint, mortar, fungus, and other contaminants the painter must scrub the surface with a brush and detergent or harsher chemicals. In some cases a wire brush or sandblasting is required.

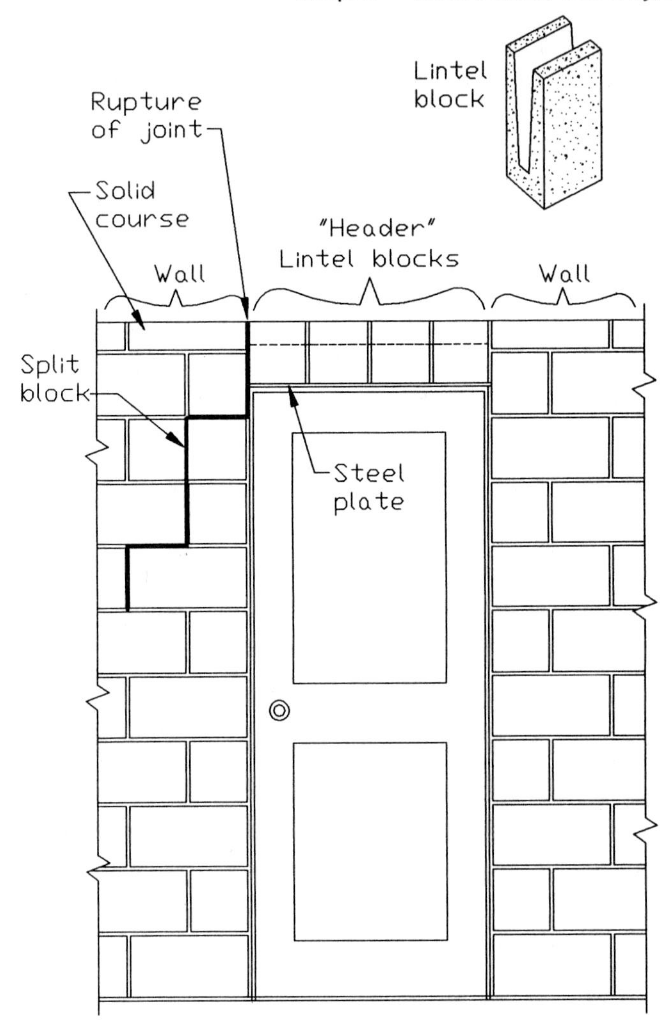

Figure 3-6 Block wall construction; poured lintel header and crack between header and wall.

The conditioner provides a bond between the old painted surface and the new top-coat paint. In every case, make a test by applying the conditioner to several places and checking for bonding. If all is well, proceed with the entire job. For best results, work in temperatures above 50°F, or preferably 70°F, and a relative humidity around 50% to 60%. The weather should permit drying times of 4 hours or slightly longer.

Concluding comments. We have examined problems with brick walls, stuccoed walls, and cement block walls. We have discussed the types of problems contractors will have to restore the buildings to their original conditions. We have discussed several types of contracts that are fair to both the owner and contractor. The objectives of the chapter have been covered. The examples used in the generic projects are actual and factual. The solutions are all workable. All the work in this chapter needs to be done by professionals. Homeowners with considerable skills may be able to perform some of the tasks, but there may be an element of risk. If the reader fits this description, proceed with caution, buy the best materials, and call for help if required.

CHAPTER SUMMARY

Brick veneer, masonry (blocks) and stucco are materials that require complex skills and great experience as well as considerable knowledge of chemical actions of mortar, cement and stucco. Although the tasks appear simple to perform, they are not. Lines must be maintained, cementing materials consistency is critical, and timing is very important. In addition, the materials are heavy and their applications require the use of scaffolds to reach the upper courses of brick and block walls. Scaffolds are also needed for applying stucco. Quality materials, workmanship, and a fair contract are the major ingredients needed to restore the outside of the home.

4

EXTERIOR WALL STRUCTURES

OBJECTIVES

To examine the construction of exterior wood-framed walls.

To examine the structure of block walls.

To understand how contractors or builders restore exterior walls.

OPENING COMMENTS

Conditions or circumstances that require corrective actions. Most of the time, homes made with framing or block wall and framing require few if any corrective actions through the life cycle. However, there are several circumstances when repairs or restorations need to be performed. Some of these situations include damage from tornadoes, hurricanes, severe winds, earthquakes, and insects such as termites and rot from water causes. Since the walls are usually covered on the inside and outside, we often cannot detect damage from insects and water until either manifests itself visually. On the other hand, we can readily see damage to the framing and block walls from violent weather or natural catastrophe.

Contractor responsibility. The contractor who is called to the site of a home that has sustained severe damage has to be knowledgeable in home construction. He or she will need to assess the physical damage and determine its extent. The assessment must take into consideration some of the following:

1. Magnitude of an infestation of termites.
2. Extent of rot.
3. Damage to sole plates, studs, and ceiling plates.
4. Damage to headers and window and door jacks.
5. Alignment of the wall for both plumb and angle.
6. Sagging of the roof or gable end from collapse of the wall studs.
7. Type of construction (western platform or balloon).
8. Extent of block-wall rupture.
9. Extent of sag of the floor joists, sills, and headers.
10. Structure of the block walls.
11. Use of lentils and reinforcing techniques.
12. Methods by which sills are anchored to block foundation walls.
13. Methods of waterproofing the exterior surface of a block wall.
14. Techniques of applying plaster or sealants to wall surfaces.

For each of these items, the contractor needs to thoroughly determine the full extent of damage by making several appropriate tests. In the order of the list, first probe the wood along the path that is visible and beyond. Illustrate to the owner where the problem stops and then discuss the techniques used to rectify the problem. If rot is evident, its cause must also be defined to the owner. Was the rot from leakage around the window or door, through the siding, or from a vent through the roof. Replacement of the rotted pieces is important, but so is eliminating the cause.

In violent storms, walls are ripped apart as if a huge ball had crashed through. Pieces are rooted, twisted, cracked, and torn out by the winds and flying debris. When walls are ripped apart, the damage usually extends to the framing around the windows and doors. In the case of hurricanes, entire walls move and may even move out of alignment. This movement causes doors and windows to stick or freeze shut. Sometimes house jacks and come-alongs need to be used to straighten the walls before permanent repairs can be made to the framing.

When studs have been ripped out of a wall section on the roof-bearing wall, the plates cannot hold up the roof or may partially hold the roof from collapsing. The roof sags. House jacks must be used to realign the roof and then new studs are installed. The contractor must perform this work.

If flooding or an earthquake damaged the foundation and basement walls, whole sections are sometimes crumbled or pushed in. Since the floor joists, framing, and flooring bear on the blocks, these parts may either sag or cave in. Here the responsibility of the contractor is to raise the floor assembly and then replace the block wall.

Homeowner's expectations. The homeowner is extremely upset from viewing the damage. Rightly so, he or she expects a large repair bill. But, in all likelihood, the actual cost may be as small as $250.00. The homeowner's insurance policy may cover the remainder. (This depends on the adjuster and the specific orders he or she has received from the home office regarding assessment of damage and optional settlement.) Therefore, the owner is much more interested in how soon the damage can be repaired and his or her life returned to normal.

The contractor understands this need and usually provides a shorter, rather than longer, time frame for restoration. The discussion between the two parties involves an explanation of the sequence of actions that will take place to effect the repairs. This explanation includes the alternatives of employing three or four skilled workers, versus one or two, to speed the work.

The owner also expects the repairs to restore the house to full strength and quality. This means that the framing and other damaged materials such as basement walls need to be equal in quality to the ones destroyed. Furthermore, where the damage was from insects or water seepage, the cause must also be removed.

Scope or types of projects to solve. For our generic projects, we will develop a problem of violent storm damage where a section of the upstairs wall in a two-story house has been destroyed by flying debris and 100+ mile-per-hour hurricane winds. The second project will detail the replacement of termite-ridden corner studs and plates. A third problem for us to study will be the restoration of a basement wall caved in by flooding.

PROJECT 1. HURRICANE-DAMAGED EXTERIOR WOOD-FRAMED WALL

Subcategories include balloon framing; western or platform framing; damaged joist assemblies; damaged plates, headers, and sills; damaged window and door units; damaged corner assemblies; out-of-plumb and square walls, sagging roof line; damaged cornice.

Primary Discussion with the Owner

Problem facing the owner. A hurricane is frightening and leaves the owner in a state of shock in its aftermath. Walls torn out or damaged from falling trees and flying debris and pressure blow-outs from accompanying tornadoes leave wide-open gaps between inside and outside. Owners faced with this view rely on family and friends to overcome the emotion of the situation by using plastic sheet vinyl to close up the gap, as if applying a band aid to an open wound.

Shortly after the storm abates, the owner will contact the insurance company and often request the name of a reliable contractor. Or he or she may personally know a contractor or may use one suggested by a friend or neighbor.

When the contractor arrives on the site and makes an inspection, the first thing done is to close up the damaged area to prevent looting and to make sure that the structure is safe to inhabit. The owner needs to be there to understand firsthand what the contractor finds and explains. In the case of an exterior wall with a section destroyed, we are faced with a bearing wall that used to carry a load of roof or gable end and part of the ceiling joist assembly. Now that support is gone. If left as is, further damage will likely result since the weight continues to bear down. Some shoring must be done immediately. Figure 4-1 shows a long section of an exterior framed wall. If several studs are pulled out, the plates will sag. If the corner studs and several wall studs are gone, too, the corner of the plate will sag. If the damage extends into a window or door, then not only studs but jacks, cripples, and headers are damaged.

In our first project, we are not concerned with theft, because the damage is on the second floor of the home. However, we are concerned with further damage to the interior of the home from lingering winds and squalls. In addition, we are also concerned for the safety of the family. We must close the opening to prevent someone from falling out and being injured. We must also be concerned with the roof sagging.

Alternatives to the problem's solution. The alternatives to a severely damaged framed wall are few. As mentioned, we need to ensure that the roof and ceiling assemblies do not sustain further and permanent damage. To do this, we would employ house jacks and several studs or 4 × 4's. The contractor must attempt to seal off the opening with plywood as a temporary fix until a full assessment can be made. The owner should be informed of these costs and agree either to pay for them or make them a part of a larger contract, which will cover the entire project.

The contractor should make no attempt to reinstall drywall or siding or replace carpeting, wood flooring, and the like. The ceiling, if damaged, should be removed to enhance safety.

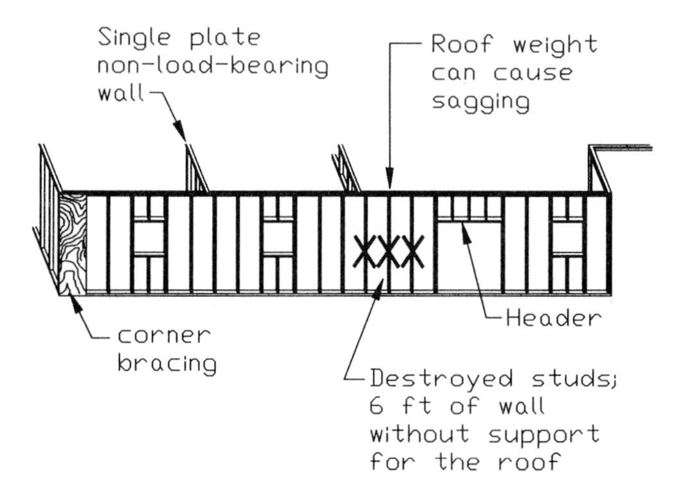

Figure 4-1 Wall framing.

In the restoration effort the contractor must replace the damaged members of the wall and framing for doors and windows and install appropriate bracing. These are standard practices. The effort may vary considerably. If the house is very old, the construction will have been similar to that shown in Figure 4-2, where oversized studs, by today's standards, were used. The siding was nailed directly to the studs, and wood lath was nailed directly to the inside surface of the studs. Corner braces were used.

If the house is two story and balloon framed, as shown in Figure 4-3, studs 16 to 18 ft long were used. Certain special details were used to support the joists. Bracing was also different. The repairs to a house of this structure will require a different approach than with western framing.

If the house was built in the early 1900s, then 3/4-in. tongue and groove sheathing was installed on a 45° angle over the outside of the studs. The inside would still have wood lath and plaster. Later in the 1900s, plywood largely replaced tongue and groove sheathing. It was quicker and simpler to install, thus saving worker-hours. Still later, plywood was used on the outside corners of the house to stabilize the frame and keep it plumb. The rest of the studs were covered with insulation panels 1/2 in. thick. In some energy-efficient structures built today, metal corner braces are nailed over the studs, and thick insulation panels up to 1½ in. thick are nailed to studs.

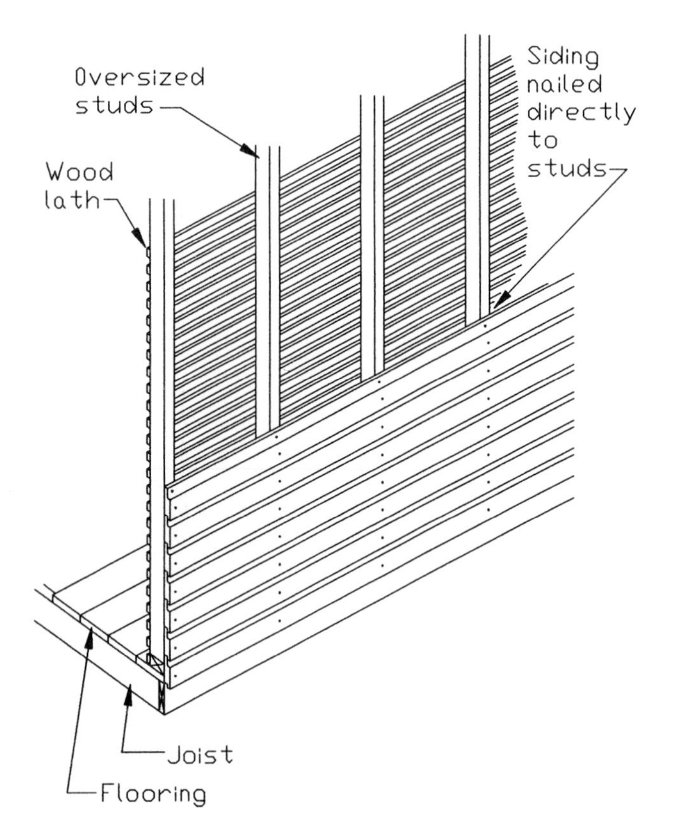

Figure 4-2 Old-time wall; oversized studs, lath, and no sheathing.

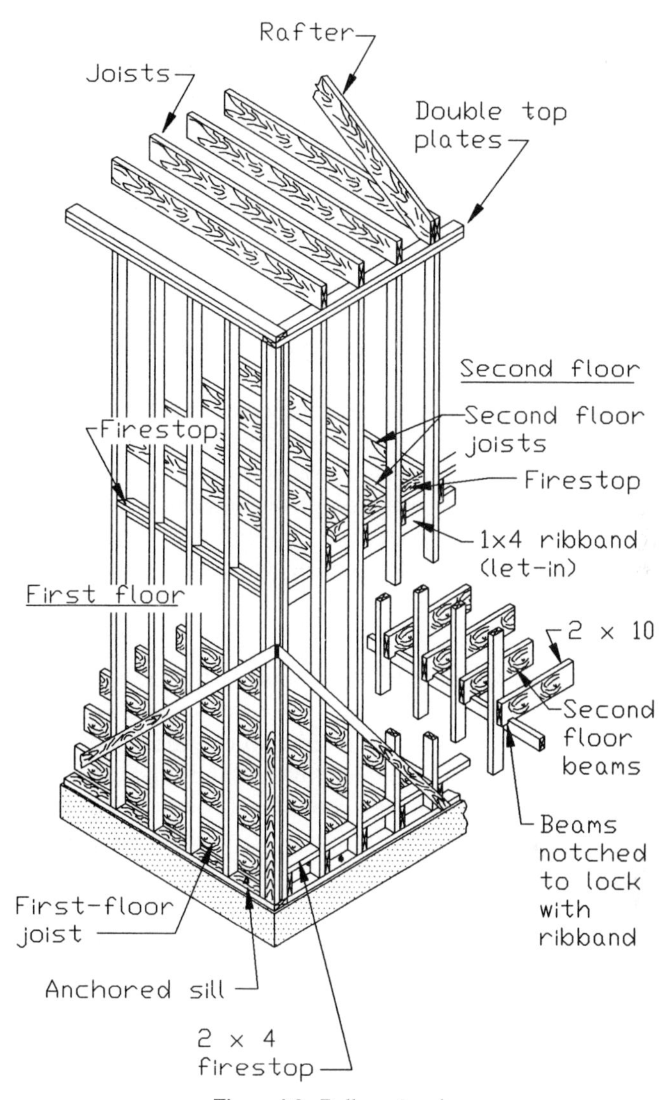

Figure 4-3 Balloon framing.

Since the wall's construction is exposed because of the damage from the storm, the style or construction technique is easily identified. This, simplifies the decisions the contractor must make with regard to restoring the wall.

Other problems facing the contractor are the need to terminate all electric service damaged, all plumbing pipes and vents pulled apart or severely dislodged, and TV cables strung through the wall. For our project, the contractor will also require scaffolding.

Statement of work and the planning effort. Normally the statements of work would be concerned with the full restoration. But, for this study, we are limiting them to the structure of the wall's framing destroyed by the storm. The exterior coverings were covered in earlier chapters. One of the first statements of work is *make necessary emergency repairs to prevent further damage, eliminate the cause for fires, prevent loss of water from ruptured pipes, increase the safety at the point of damage, and reduce the opportunity for theft and vandalism.* Then the repair effort can be stated something like this: *Remove and replace all damaged framing materials. If required, remove siding, sheathing, and interior wall coverings to expose the full extent of damage.* Inherent in the replacement effort is the need to reestablish the lines on the wall joists and roof. A statement of work would be to *ensure that the walls after being replaced are plumb and in line with the rest of the wall, verify the accuracy of the ceiling joist alignment, and ensure the accurate alignment of the roof overhang and cornice or gable end.*

Since emergency repairs are a part of this project, we have very little time to plan for those repairs. First, we must rely on our knowledge and skill to avert further damage. Second, we need to satisfy the owner's desire to secure the property.

Beyond this effort, we have a short time to plan for the restoration. We should spend some time carefully examining the damage because frequently the fierce winds lift the walls and the roof from the walls. Following a thorough examination, we must plan to estimate the job, obtain contracts, and provide the materials and workers if we win the contract.

Side Note to Owner and Contractor. Because of the conditions following a violent storm, we can expect that all contractors and subcontractors will be heavily engaged. Therefore, we will satisfy this owner and also attempt to secure other work. To do this, we must meet with many owners and solicit their job offers. Then we need to quickly provide bids and sign contracts. In some cases a handshake will do, but it is not advised. Our bids must be competitive or we stand a real chance of losing the work.

We also must be sure that materials are available in sufficient quantities since there were only finite quantities available before the storm at the usual prices. Materials brought in after the storm are out-of-cycle purchases and thus incur added costs. This accounts, in part, for the elevated contract prices associated with a storm restoration time frame.

The subcontractor electrician and plumber, if required, must also submit contract bids for the repairs.

Our office and other overhead personnel can expect to work long hours to meet this emergency need. The situation is an excellent one to help meet our goals for covering fixed costs and making a profit. If the office is equipped with desktop computers and a variety of software used in estimating and billing, we can create excellent contracts from boiler plates. Our database can be used to provide cost estimates for each phase of the effort, and other entries can provide pricing at the usual rates. We will have to verify the prices, but a few calls will check them out. This means that we can offer the owner a contract bid in several hours to one day for small jobs and in several days for large jobs, where many walls were destroyed. Speed is important.

For this project, we can expect to offer the owner a bid/contract offer in a day or two at the most.

Contract. For our generic project, the contract could be either a time and material type if there are many unknowns, such as moved walls, resecuring walls to the foundation, reestablishing plumb lines and level lines, and the like, or the contract can be a fixed-price type when careful examination has revealed no hidden or unexpected work.

The body of the contract must reiterate the statements of work in narrative form. This clearly states the work to be done. The office copy of the contract may include the price out for direct and indirect labor, fixed and variable overhead, and allowance for profit. After the contract is won and proceeds from the owner have been received, the breakdown is applied to the various accounts in the computer.

Normally, we would expect several contractors to submit bids for their specialty work. These would be siding, electrical, plumbing, heating and air conditioning, painting, drywall, and carpet or floor finishing. But we are only dealing with the framing aspects, so carpenters will perform the work.

Material Assessment

Direct materials	Uses/purposes
2 × 4's	Studs and braces, soles and plates
2 × 8's, 10's, or 12's	Headers
Common nails	Securing the studs, plate, and headers

Indirect materials	Uses/purposes
Plywood	Secure the property from pillage
Nails	To hold plywood in place
2 × 4's or 4 × 4's	Temporary bracing and shoring

Support materials	Uses/purposes
Carpentry tools	Construction
Come-a-longs	Restore plumb and level lines
House jacks	Raise plates, joist assemblies, and roof units
Scaffolds and planks	Work platforms
Power saws and cords	Eases work effort

Outside contractor support	Uses/purposes
None scheduled	

Activities Planning Chart

Activities	Time line (days)						
	1	2	3	4	5	6	7
1. Emergency repairs	×						
2. Contract development		×					

(*continued*)

Activities	1	2	3	4	5	6	7
3. Removal of damaged materials	___	___	X	X	___	___	___
4. Restoration of wall, joist, and roof lines	___	___	___	X	X	___	___
5. Installation of wall parts	___	___	___	___	X	X	___

Time line (days)

Reconstruction

Emergency repairs. We have already identified the purpose for emergency repairs and some of the reasons for making them. In our generic project, we need to provide immediate support for the ceiling joists that lay on the wall plates. We also need to shore up the roof, which also rests on the wall. So, we rent 6-ft telescoping house jacks and install them in several places along the damaged wall. Along the top of the wall, we use doubled 2 × 4's or 4 × 4 pieces to spread the downward force and thereby reestablish the ceiling line and roof line at the same time.

While installing the jacks we are careful to keep them from extending out past the wall line. We take this precaution since the outside of the wall needs to be covered. But before we can install the plywood and chipboard sheathing, we must cut and nail several temporary studs as well as a corner in place. Cutting and nailing the plywood is a routine job. Over the plywood, we apply 15-lb felt. However, the sheathing alone will prevent rain from entering the building.

Other problems to solve on a temporary basis include electrical and plumbing services that were disrupted. Electrical outlets and switches torn loose need to be terminated before the power is restored.

Contract development. We have simplified the contract development to only the part performed by the carpenters to restore the wall's structure. We have mentioned the wide variety of contractor specialists that would ultimately be involved in the full restoration. For the general contractor to manage the full restoration, all the different specialists would submit contract bids. Then a single bid from the general contractor to the owner would be the vehicle to obtain the work.

The general contractor could permit each subcontractor to supply the needed materials or he or she could supply all needed materials and require only labor cost bids.

For our generic project, we will have a simple contract with a cost for emergency repairs separated from final repairs to the wall.

Removal of damaged materials. This part of the job is dangerous. Considerable care must be used to avoid mishap. Generally, we use the following routine:

1. Remove the temporary sheathing.

2. Remove the siding and sheathing over the damaged area and back into the undamaged area as far as required.

3. Move the house jacks inside the room to free up the wall plates. We use a second pair of jacks to do this.

4. Shore up the roof line to take the weight off the wall plates.

5. Remove the interior wall coverings and moldings to expose the entire damaged area.

6. Move back the carpeting if there is any.

7. Remove the studs, window framing, door framing, corner units, plates, and sole plates if they are damaged.

After these items are removed, further damage requiring additional work must be carefully evaluated by close examination. Such items as building true, anchoring to the foundation, and whether members of the floor and subflooring are soundly nailed to each other are some of the first efforts.

When we find evidence that these members have been disturbed, we effect repairs accordingly. Then we test for plumb, square, and alignment. We use the plumb bob to verify that the walls are plumb. We use the square and right triangle (3-4-5) to verify squareness at the corners. We use a mason line to verify alignment of the wall. The plumb bob will tell us how much out of plumb the wall and corner are. The square will tell us if the house has been racked and by how much. The line will tell us if the wall is bulged out or caved in and by how much. We can then decide what steps to take. Since houses are frequently out of true measure by 1/4 to 1/2 in., we can advise the owner accordingly. He or she might be willing to accept these tolerances since they will not appreciably affect the looks or integrity of the structure.

Restoration of the ceiling and roof line. These two aspects are vital and need to be done before the wall members are replaced. To realign the joist assembly of the ceiling, we employ 4 × 4's and house jacks. Recall that we moved the jacks off the wall to remove the damaged materials. When we moved them into the building, we temporarily supported the roof with 2 × 4's from ground to under the cornice.

Now we concentrate on ensuring that the joists are in alignment and are level. This is done with the house jacks. Normally, we would take precautions to avoid further damage to the ceiling; but if it is already damaged, marks made by the 4 × 4's under pressure will not appreciably add to the repair. However, should the ceiling have survived without damage, we will use either 1-ft-wide plywood strips between the 4 × 4's and ceiling or pad the 4 × 4's with carpet strips.

We can check the ceiling's accuracy with level and line. By stretching the line across the ceiling perpendicular to the run of the joists, we can check and correct sags and high spots. Then we can check for level along the run of a joist. This task may take 1 hour to 1 day depending on the size of the room and the extent of damage.

We will delay final realignment of the roof until after the wall has been restored. This is usual since the roof bears on the wall. However, once the wall is restored as described below, we remove the temporary 2 × 4's holding it up and make adjustments or modifications where the roof and cornice were damaged. (See Chapter 1 for this detail.)

Restoration of the wall framing. The next part of the job is to replace the materials removed. For our generic project, we are dealing with a western framed house. This is the simplest type to repair. However, certain changes have been made to the dimensions of studs and 2 × 6 and wider materials recently. Their dimensions have been made slightly smaller. We may need to pad or add shims to the studs if a difference exists. With regard to headers, we would simply use thicker spacers between members to account for the difference in thicknesses.

Installing all the new members may require that each be installed separately, rather than making a wall assembly and raising it in place. This is not difficult, but requires slightly more time. All nailing will be toenail type, except for nailing headers and jack studs in window and door openings and manufacturing a new corner assembly.

We will match the stud spacing of the original wall. This could be either 16 in. o.c. (usual) or 24 in. o.c. If the house is very old, some other spacing might have been used.

Concluding comments. This type of restoration has many aspects that require the skill and knowledge of craftspeople. These are not simple tasks. Great care, patience, and attention to the smallest detail are required to avoid total catastrophe. We have examined the sequence needed to restore the wall and associated joist assembly and roof assembly. The contractor knows that these tasks are needed and require experts. He or she knows that skilled labor costs premium dollars. The owner needs to understand this, too. Unskilled carpenters or novice builders, although less costly, may, in fact, create further damage or at best perform a shoddy restoration that will have to be redone again.

PROJECT 2. RESTORATION OF A WOOD-FRAMED WALL DAMAGED BY TERMITES

Subcategories include damage from dry-rot, ants, rodents, fungus, standing water, dampness, lack of ventilation, improperly cured lumber, and infected wood installed during the building of the house.

Primary Discussion with the Owner

Problem facing the owner. When the contractor was called to the site of the termite-infected framing, he or she could tell immediately that the problem had been present for some time. The exterminator had made an inspection of the foundation and found evidence of more than one colony of termites. The damage seemed to extend from the corner of the house in both directions. The owner had the exterminator remove some of the siding to expose the wall framing and sheeting. The damage extended to the wall plates and into sole plates and wall studs. The sheathing was also infected.

The exterminator recommended spraying in the ground along the walls on both the outside and inside of the building. Further spraying would need to be done when the damaged materials were removed.

The owner was facing a severe problem since the following parts of the framing were key to the building's integrity and strength:

1. Sill
2. Joists
3. Subfloor
4. Sole plates
5. Wall and corner studs
6. Wall double plates
7. Ceiling joists

Each of these plays an important role in the house's strength and stability. The sill, if damaged, cannot support the floor joist assembly properly, and the floor would squeak and feel spongy. The wall assembly, consisting of sole plate, studs, and ceiling plates, provides the bearing surface for the ceiling joists and roof rafters. When damaged and weakened by insects who eat wood, the structure cannot support the weight. Thus the weight bears down on the sheathing, drywall, and siding to some extent. The idea of total collapse is not viable in most situations, but sagging and structural weakness are.

Alternatives to the problem's solution. The owner and contractor must first eliminate the termite nests. Then decisions must be made about the structural members. Should they be removed and replaced or should new members be spliced alongside the damaged ones? In almost every situation of severe damage of this type, the owner and contractor come to an agreement on replacing the members that are destroyed completely and those that are in the bearing wall. Other members that were infected and partially damaged, but are now free of insects, are spliced with new members to restore the soundness of the floor, ceiling, and wall assemblies.

Generally, this is the sound approach. However, the contractor should advise the owner that the nominal cost of a few studs or piece of sheathing should not be the deciding factor. Rather, it would be foolish to save the damaged studs during the work effort in removing the sole plates. By the same token, it would be foolish to replace a 12-ft joist when just the first 3 to 4 ft have been infected. A splice would certainly restore the soundness of the floor.

Figure 4-4 is a view of the normal structure where the damage in this problem exists. It shows the relationship of the members and those that bear onto others.

Statement of work and the planning effort. The work of replacing or shoring up the members of the wall and floor assemblies is relatively simple and straightforward for any carpenter. The statement of work is *determine the extent and*

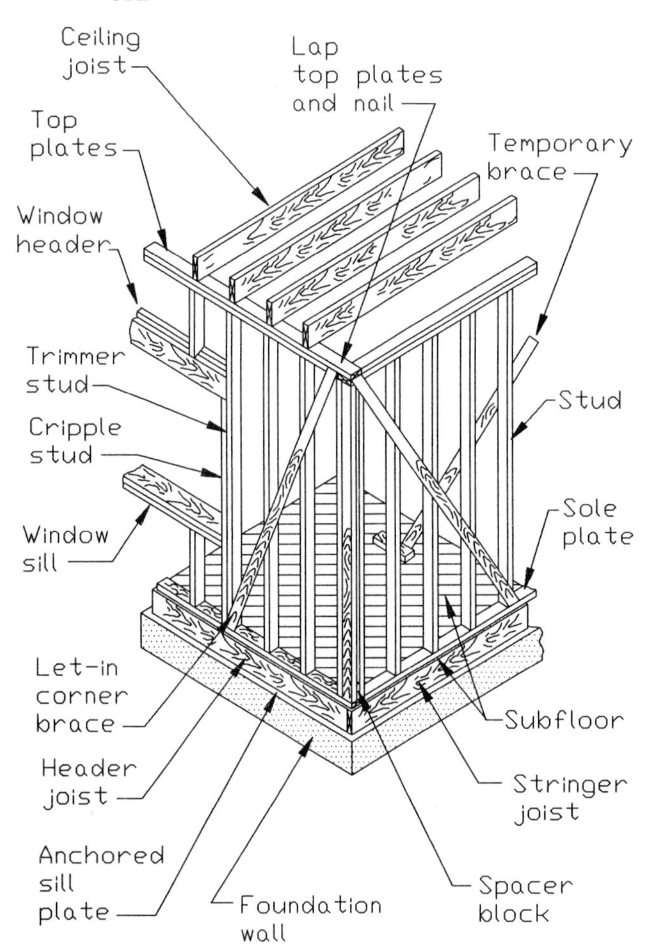

Figure 4-4 Corner of platform framing.

range of damage caused by termites and remove the pieces that bear a load. For the members that support the floor, we *splice the joists after the primary joists have been fully treated with chemicals.* We would also expect to replace damaged sheathing and subflooring. The statement of work would read *replace all damaged sheathing and subflooring beyond the visible signs of damage or infestation.*

The planning effort requires preparing a contract, assigning someone to perform the work, and ensuring that the termites and eggs have been eliminated. We also want to schedule the job during a spell of dry weather to preclude causing added damage. The estimate of materials should be easily defined and materials should be readily available locally.

Some shoring of the wall and ceiling joists will probably be needed. So we should plan to use house jacks to support these weights during repairs.

Contract. The contract for this repair effort could be a fixed price, fixed-price labor and variable materials, or time and materials. The type of contract we used depends on our capacity to accurately assess the amount of damage and its true extent. If the true extent is observable, then the fixed-price contract would benefit the owner and us as well. But if there is even a small factor of unknown, say 10%, then we suggest a time and materials contract, and if the owner balked, we would modify the type to a fixed labor and variable materials. The owner in this case might spend more for the restoration due to the need for us to make an allowance for the unknown in our labor figure.

The body of the contract would contain statements pertaining to the statements of work and the price-out. For example, it could look like this:

> The infected materials that cannot perform their intended purposes will be removed and replaced with original quality materials. Due to the unknown extent of damage to some areas, the costs of materials with markup will be priced out after the job is concluded. Materials that are damaged but are clean of infestation shall be spliced to restore the member to original purpose and durability. Where materials are in contact with masonry or subject to moisture, treated lumber will be used. The area of infestation will be treated for termites before the siding is reapplied. The estimate for materials is $×××.××.
>
> This contract includes termite protection.
>
> Labor and termite protection costs are fixed at $××××.××.
>
> The total estimated cost for the restoration is $××××.××.

Material Assessment

Direct materials	**Uses/purposes**
Studs	Replace and splice wall studs and plates
2 × 8's	Sill and floor joist materials
Plywood (or substitute)	Wall sheathing and subflooring
Nails	Secure wood
Felt	Applied to outside of sheathing
Caulking	Seal the new sill to the foundation

Indirect materials	**Uses/purposes**
2 × 4's or 4 × 4's	T or top pieces supporting joists when used with house jacks
Plastic sheet (6 mil)	Protect the exposed wall area when left open

Support materials	**Uses/purposes**
Carpentry tools	Construction
Power saws	Eases the work
House jacks	Support joists and realign joist assemblies
Ladders	Needed for reaching upper wall areas and ceiling joists

Outside contractor support	**Uses/purposes**
Termite protection subcontractor	Treat the new work and retreat the surrounding area and remainder of the perimeter

Activities Planning Chart

Activities	Time line (days) 1	2	3	4	5	6	7
1. Site examination and estimation	X						
2. Contract preparation and scheduling		X					
3. Materials delivery and restoration			X	X	X	X	
4. Termite treatment							X

Reconstruction

Site examination and estimation. We have discussed the need to make a thorough site examination. The need was clear and influenced the type of contract we offered the owner and helped us arrive at a fair estimate of the work required. This part of the reconstruction may have required us to remove more of the materials, such as sheathing or dry wall, to make a full assessment. Even in this effort we may not have been able to completely identify everything that was damaged. Therefore, our estimate had to include the variable on materials costs, and we made a small allowance for labor in our fixed price.

The estimator from the contracting company, along with the termite protection employee and owner, probed the floor, joists, sill, header, and sheathing for several feet beyond the obviously damaged areas to be sure to arrive at a fair understanding of the work required to restore the building. The estimator made notes about the parts that had to be removed and replaced and those that could be spliced. He or she also determined where temporary shoring had to be employed to realign the joists, ceilings, and floors for the purpose of defining the required materials.

Contract preparation and scheduling. The estimator or contractor converted the on-site examination and notes into the contract/bid. For his or her part, the contract allowed for direct and indirect labor, an allowance for fixed overhead, an allowance for profit, and an approximation of the material costs, plus the variable costs associated with the jobs, such as insurance, transportation costs for materials, and others.

The office copy of the contract should contain a work sheet or carbon page that indicates the direct and indirect costs for an audit trail.

Due to the nature of the work, one or at the most two carpenters are assigned to the repair effort. The schedule needs to be set up with the owner since access to the house is required, and therefore someone in the owner's family should be home. We could schedule the work to be done in almost any type of weather, but weather that includes rain or snow should be avoided if possible.

As our activity chart shows, we plan 4 days of carpentry work to restore the framing and replace the sheathing. The actual time will vary with each job, of course.

Materials delivery and restoration. The carpenters will drive to the lumberyard on the first day of their schedule and pick up the materials if the materials were not delivered the day before work was scheduled to begin. Since the estimator made careful notes, these were translated to actual materials lists like the materials assessment, but they are more complete with actual quantities, sizes and types (treated and untreated, grades, and so on).

The carpenters ensure the stability of the floor and ceiling joist assemblies by installing the jacks before removing the damaged stud walls, plates, headers, and sills. The work would follow this approximate schedule:

1. Remove the interior wall covering. This could be drywall, lath and plaster, wainscot, or paneling. The amount removed should be about the distance from the corner to the expected length of damage on each wall in our generic problem.

2. Remove exterior wall and cornice materials. The carpenters go outside and remove some of the cornice to expose the wall plates and tops of studs. They remove siding and sheathing not yet removed during the inspection phase.

3. Examine the floor joists. The carpenters place the house jacks such that the joists bearing on the sill are at zero pressure or pounds per square inch. Since the joists

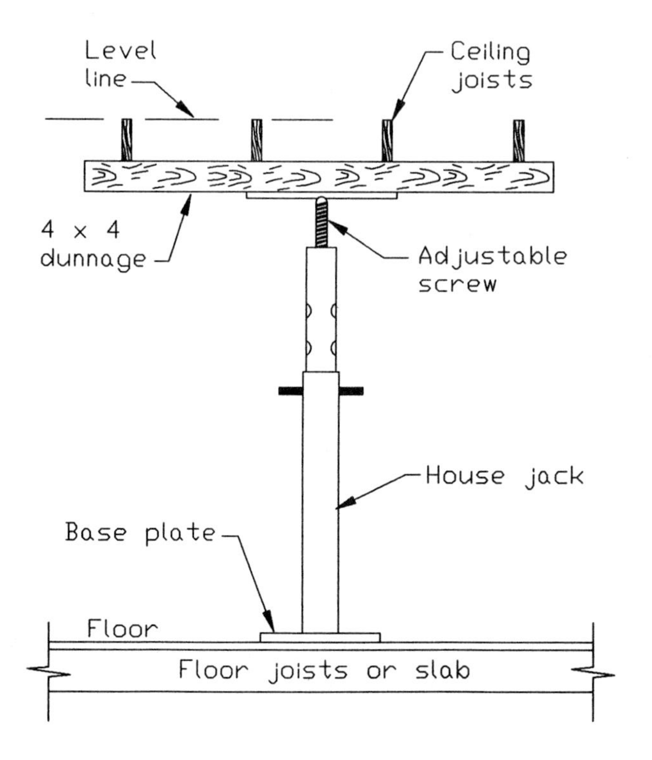

Figure 4-5 House jacks under joists.

only run in one direction, only one pair of jacks is needed for this job. Figure 4-5 shows what this looks like.

4. Shore up the corner and roof. From the outside, the carpenters install a doubled 2 × 4 from the ground to the wall plate to transfer the weight of the ceiling joists and rafters to the 2 × 4.

5. Remove the studs and plates. The carpenters remove the damaged studs and corner assembly. Then they cut through the sole and ceiling wall plates to remove the damaged members.

6. Sill, header, and joist removal. With the wall members out of the way, the carpenters will begin to eliminate the damaged sills, headers, and joists that form the outside of the joist assembly. The headers are cut off past the point of damage. The sills are also cut off at these points. The result is an exposed foundation and joist ends on three or more floor joists.

7. Joist examination. With the joists exposed, we determine the need to replace or splice the joists. If removal is elected, then we must make arrangements to free the opposite end from the girder or other wall. If splicing is elected, then we should trim away the badly damaged area or ends. We need to delay installing the splices for now.

8. Subfloor inspection and removal. If inspection reveals damage to the subfloor, and the floor is carpeted or tiled or has a hard wood flooring, we must remove these first. To pull the carpet back is simple; 5 minutes and we are done. To deal with vinyl tile, we must sacrifice the tile. To deal with hardwood flooring, we must remove the baseboards and shoe moldings; then we take up the flooring while trying to save as much as possible. After the subflooring is exposed, we simply cut through it with a power saw and remove it.

The restoration is almost the reverse of the removal process. We replace the sills first. These are fastened to the foundation with the bolts embedded in the concrete and toenailed into the adjoining sills. Then the joists should be spliced by bolting new pieces to the old ones and also nailing them together. The next tasks are to replace the header joist, subflooring, wall sole plates, corner wall assembly, studs, and sheathing. If there was bracing in the corner, this is replaced before the sheathing is installed.

If the infestation and damage extended to the ceiling joists and wall plates, we will need to install jacks on the floor and support the ceiling before removing the wall plates and splicing the joists.

Termite protection. The exterminator must return to the job site to complete the protection. This includes impregnating the new wood with chemicals to reduce the recurrence of the problem. The ground below the wall also needs protection to kill termite nests.

Concluding comments. This problem usually requires a costly solution. We have only examined the restoration of the framing part of the work. The other parts, both interior and exterior, must be restored as well. Considerable skill and understanding about home construction are necessary to perform this work. An owner must have these skills and knowledge before attempting this work. Even if there is understanding after studying this problem solution and adapting it to the actual conditions in your home, it still might be better to use the expertise of a contractor.

PROJECT 3. RESTORATION OF DESTROYED CEMENT-BLOCK WALLS

Subcategories include restoration of the basic block wall; built-in pilasters; stacked block versus running bond; reinforced headers; installation of cap block; installation of rebars; channel blocks; window sills; reinforced openings for various purposes.

Primary Concerns of the Owner

Problem facing the owner. The owner returns to his or her home after a major storm and finds the block wall on one side of the house caved in or blown out. If the storm was a tornado, the likely damage will be a blown-out wall due to the severe change in air pressure outside the house. If the damage was caused by a flood or hurricane, the wall will most likely have caved inward. In either situation the owner has a very serious problem. Any framing above the wall, such as joist assembly or roof, or gable end will have either caved in as well or is hanging free and may be ready to fall.

Two likely scenarios are present. One, any work to be done must begin with making the work area safe. Damaged but relatively stable assemblies must be shored before cleanup of debris begins. Jacks, 2 × 4's, and other dunnage need to be used to hold assemblies in place. At this stage of the task, accuracy of alignment is not essential. That can be done later. Stability is paramount.

The second scenario is when the debris is removed from the job site. Once again, safety is critical. Blocks or rubble blocks need to be cleaned away from the site to permit a critical assessment of the extent of damage. Some of the damaged wall may have been a pilaster or rebar-reinforced block acting like a pilaster. Other parts of the wall may have had windows installed. Still other parts may have been the corners. In each case, we must assess the total effect of the structure.

In all probability we will have to remove more of the blocks. Some will be split and others that are part of the pilaster may have been laid over, and we will have to rebuild from the footing up. Damaged block around windows will need to have the header removed and later rebuilt.

The costs for restoring the walls from this kind of damage may be moderately expensive to very expensive. For example, the total square feet of wall to replace is a factor. The type of block and installation technique are factors relative to cost. The number of pilasters to rebuild adds to the cost. And the finish applied to the walls also varies greatly and thus has a direct bearing on the cost.

The owner who has the skills to lay block may be able to restore the damaged walls. However, the tasks are not simple and some of the work must be done from a scaffold. Mason contractors employ laborers to mix the mortar, lay up the block so that the mason does not have to walk to get them, and so on. The mason's skill will ensure that the rows and joints are even and finished. Finally, the urgent need to restore the building may also be a factor in determining whether the homeowner makes the repairs or contracts for the work.

Alternatives to the problem's solution. There are several alternatives that the owner can elect. The first is to delay restoration until the best price alternative can be obtained. In this case the caved-in wall can be temporarily closed up with plywood. The remainder of the house is either secure and livable, or other damage has made the entire place unlivable.

The insurance adjuster must assess the property before much of the cleanup begins, but not necessarily before the temporary shoring is performed. All costs associated with the restoration may be covered by the homeowner's policy.

With regard to the work that needs to be done, the owner must have the house restored to its original configuration. This means that the same type and kind of block must be used. Steel rebars need to be embedded in pilasters and headers. All joints must be struck to match the rest of the walls. Usually, two patterns of block laying are used on residential homes. One is the running-row style where the blocks are laid as shown in Figure 4-6a. The other is the stacked style, as shown in Figure 4-6b.

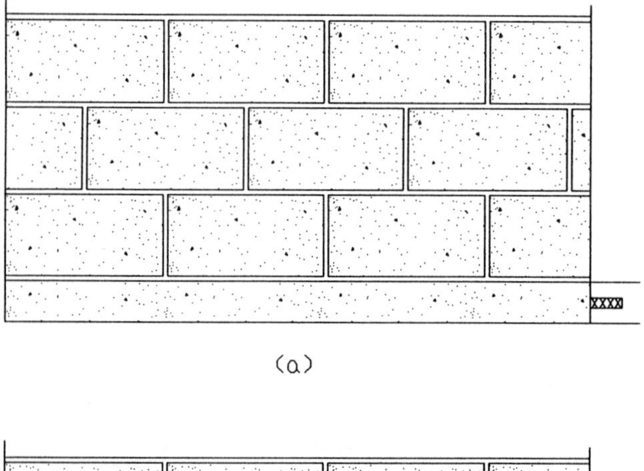

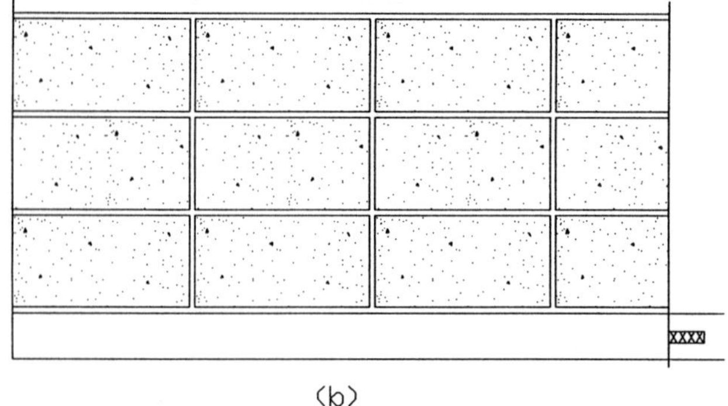

Figure 4-6 Running course and stacked courses.

For our generic project, we assume a 20-ft-long wall with one pilaster column built in. The wall has a header over a 3-ft window opening. The height of the wall is 8 ft 4 in. from footing to ceiling or floor joist. A 3½-in.-thick concrete slab was poured after the walls were raised.

Statement of work and the planning effort. Our work effort must result in the full restoration of the wall. Therefore the statements would be *restore the block wall to its original condition to include the following specifications:*

1. Cement blocks will be used.

2. The style or pattern of laying the blocks will duplicate the original wall.

3. The header over the window will be reinforced with two ½-in.-diameter rebars spaced correctly in the concrete fill of the header blocks, or the area may be formed and concrete poured into the form with steel reinforcement.

4. Cap the wall with the cap blocks and install sill bolts on 4-ft centers.

5. Rebuild the pilaster that supports the girder.

6. Apply an expansion strip between the first course and slab floor and apply a sealer over the expansion materials.

The planning effort includes many of the usual conditions and the creation of the contracts after the estimator has determined the extent of damage. We will work with the insurance adjuster in defining and assisting with establishing costs for the restoration. We will need a time frame for repairs to begin and the duration of the work itself. The contractor will also build in time for delivery of materials and availability of workers. Finally, we will require time to obtain permits and schedule building inspector inspections.

Obtaining a fair contract. The contractor and owner must recognize that many costs are incurred when damaged structures are to be repaired. These costs are not usually present during new construction. They include the emergency repairs, the need to clean damaged materials from the job site before reconstruction begins, and the removal of some undamaged materials to prepare a better starting point for repairs.

Therefore, the body of the contract should contain several paragraphs so that the owner has a clear understanding of the work to be done and their associated costs. This separation also permits the use of different types of contract styles. For example, in this project, a flat-rate hourly charge would be appropriate for the emergency repairs. The materials used for these may not necessarily be used in the final repairs and thus add to the burden of costs.

The restoration work could be supported with a fixed-price contract since we are able to accurately determine the material requirements and have knowledge of work performance and hourly rates used by masons and laborers. We also know how to apportion fixed and overhead variable costs by applying company policy.

So, in developing the body of the contract for this project, we will employ two separate types to arrive at a total cost.

Material Assessment

Direct materials	Uses/purposes
Cement blocks	Walls and pilasters
Mortar mix and sand	Bond blocks together
Rebars	Reinforce headers, pilasters, and corners

Indirect materials	Uses/purposes
Plywood	Cover the opening to provide security
Vinyl sheet goods	Protect the property from rain and snow
Dunnage and 2-in. stock	Used with house jacks to support sagging joist assemblies
1 and 2-in. Stock	Form lintels

Support materials	Uses/purposes
House jacks	Support sagging joist assemblies
Mason tools	Perform work
Hoes, shovels	Mix mortar
Mortar mix machine	Eases work
Bolt cutters	Cut rebars
Power saw with masonry blade	Cut blocks to fit

Outside contractor support	Uses/purposes
None	

Activities Planning Chart

Activities	Time line (days) 1	2	3	4	5	6	7
1. Emergency repairs	X	X					
2. Estimates and contracts			X				
3. Restoration of the wall				X	X	X	
4. Replacement of joist assembly						X	
5. Job site cleanup						X	

Reconstruction

Emergency repairs. We have already described the emergency repairs in a very general way. Essentially, they prevent further damage to the house's interior and further collapse of the joist assembly, if there is such a danger.

For our generic project, we will list the tasks that could take more than 1 day.

1. Inspection of the joist assembly must be done first to determine if it will require shoring or not.

2. Shoring materials need to be obtained from the lumberyard and from our equipment stores. Two by fours and heavier lumber for crossties and house jacks would need to be brought to the site.

3. Workers will need to remove enough broken blocks to have a safe walking area to install the jacks and dunnage.

4. After the joists and walls tied to the joists are secured, closing the wall is next. For this, we apply sheets of 7/16- or 1/2-in. sheathing plywood. The installation of plywood may be sufficient, but we could also apply 6-mil-thick vinyl sheeting over the plywood.

5. If there are electrical services in the wall damaged by the storm, the electrician will need to disconnect the service to prevent fires.

6. Plumbers will need to shut off service to the pipes attached to or embedded in the damaged wall.

7. Heating and air-conditioning people will need to examine the damage done to ductwork and controls attached to the joist assembly or damaged wall.

8. If the wall interior was finished with paneling or drywall, this debris should be removed to make a safe work area and permit the workers access to service in the wall.

As the homeowner and we survey the emergency repairs, we note the extent of the effort and the variety of people used. This joint effort will help assure a fair cost of services.

Estimates and contract preparation. The contractor who handles the entire job will have the subcontractors provide separate contracts. These ideas were already covered before. But certain expenses associated with estimating and contract preparation need to be explained. In earlier chapters, we discussed some of these.

As these items pertain to this project, the estimates require measuring the length of the wall to be restored. Then we must calculate the number of blocks required for the job. Fortunately, blocks have remained the same size for years. Thus the new ones will fit into the gap in the wall as long as the mason uses the appropriate amounts of mortar between blocks and rows. The estimator will also calculate the number of bags of mortar mix and sand needed to make the mortar. If headers were made from blocks or concrete, these items will also be added to the material listing that the suppliers deliver to the job site.

The estimate for labor will include the use of laborers to clear away the damaged block. These extra activities cause the price to increase.

In addition, the time and effort by the contractor to assemble subcontracts, calculate materials, order and have materials delivered to the job site, perform other office and management tasks, and prepare the owner's final contract/bid are overhead costs that are a percentage expressed as a markup in materials or labor or both.

Restoration of the wall. Because of the need to clear the job area of damaged block and determine the plumb of the remaining wall, and other related efforts, replacement of blocks must be delayed. We may need to remove blocks all the way down to the footing and somewhat into the undamaged areas on either side of the gap in the wall as well.

However, when we are ready to install the new blocks, replace the pilaster if required and reinforce the wall with steel rods and concrete fill, and rebuild the headers, laborers will prepare the site for the masons. They will set up scaffolds and load the mortar board and blocks onto the planks on the scaffold for the mason.

Then the mason will stretch a line along either the outside or inside of the wall along the top line of the first course. With this line in place, the mason will apply two strips of fresh mortar along the footing. Next he or she will apply mortar to the tips on the end of the first block. This is called "buttering the ends." Then the mason lifts the block with two hands and sets it on the bed of mortar and in line with the stretched line. The mason then uses a pointed trowel to tap the block until it is level with the line and in line with it as well.

Before installing the second block, the mason applies or "butters" the end of the first block with fresh mortar and butters the end of the second block and places it on the wall. Once in place, the mason uses a pointed trowel and scrapes excess mortar from the joint with an upward stroke. The object is to have a smooth, even mortar that is flush with the surface of the block on the inside and outside. The mason must avoid splattering mortar onto adjacent blocks and those in the lower courses. The process is repeated as often as necessary until the course is complete.

The blocks on the second course in our project are installed in the running pattern which means that the vertical joint is offset by one-half the length of the block. On the second and other courses, the mason must scrape the excess mortar from the joints and course as well.

After several courses are laid up, a striking tool is used to compress the joints into a concave pattern. This is important because the compression adds considerably to the water-sealing properties of the wall.

Figure 4-7 shows some of the details pertaining to this project. First is the joint that has been struck. Next is the pilaster made from blocks and reinforced. Last is the example of the header made from concrete and the alternative made from lentil blocks.

The final task is to seal the outside of the wall with a coat of bituminous asphalt if below ground or to paint the wall if above ground level.

Restoration of the joist assembly. We jacked up the joist assembly just a little above what was needed to complete the wall. This was about 1/2 in. above its normal position at the most. Now, with the wall complete and allowed to set for a minimum of 24 hours, we can apply a thin coat of mortar along the top of the last course and then lower the sill into the mortar. The weight of the joist assembly will

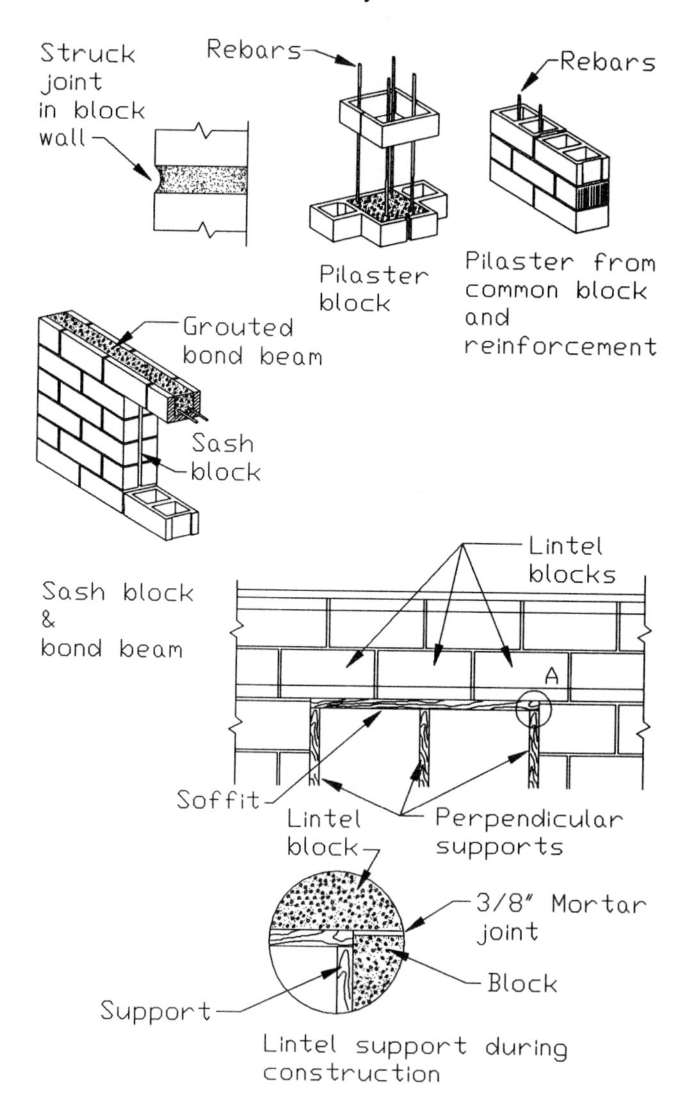

Figure 4-7 Striking joint pilaster and header.

compress the mortar and squeeze it out on both sides of the sill. We use a trowel to scrape away the excess. The mortar bed is a seal for both water and insects.

Once the sill and joists are in place, we must add nails to the assembly to remove the chance of a squeaking floor. In all likelihood, we will need to refasten the subflooring to the joists and toenail the joists onto the header or sill.

Job site cleanup. The making of mortar and concrete at the job site always creates small piles of sand and gravel, mortar droppings, and clumps of concrete and mortar. There are also broken and cut blocks, end pieces of rebar, nails, and numerous other material bits and pieces that we must clean up from the site. If the masons laid the blocks from the inside of the building, we also have to clean the floor where mortar or a slurry of water and cement got under the protective vinyl on the floor. For this job, we may need to use a mild solution of muriatic acid and water. Then we follow up with clear water and mops to dry the floor.

All dunnage and the house jacks, scaffolding and boards, mortar box, and boards and tools need to be taken out and cleaned for reuse.

Concluding comments. This project was an example of a severe situation that usually is coupled with other damage to exterior walls, floors, interior walls, and their finishings. We have only examined the walls in this discussion and descriptions. The owner will use this knowledge to determine the extent of the damage and to decide whether or not to make the repairs himself, act as the general contractor, or employ the specialties of a general contractor to have the work done. Regardless of the alternative selected, the knowledge gained from the descriptions places you in a much stronger position to appreciate the magnitude and all the nuances associated with the restoration.

CHAPTER SUMMARY

Not all damage to the house is restricted to the roof and siding. Frequently, natural causes, whether they are storm related, casual water that causes rot, or infestation, create situations where we need to replace framing materials. The repair activities are labor intensive and usually are costly since there are other products fastened to the frame. We need to have an excellent understanding of the structure of a house. We need skills that match the knowledge to minimize costs and prevent unsafe practices. We studied many different problems in the various projects, and we solved them. Some homeowners are quite capable of making the most difficult restorations. But if there is the smallest doubt about how to proceed even after studying this chapter, the only safe solution is to call a contractor. We will restore the home.

5

WINDOW AND DOOR UNITS

OBJECTIVES

To restore wooden double-hung windows (pulley and rope and tension bars).
To free sealed or painted-shut windows.
To replace old wooden windows with modern double-insulated ones and retain the architecture.
To replace aluminum single-hung window units with improved ones and retain the architecture.
To replace weatherstripping on the window.
To refit and rehang an exterior door unit.
To reinsulate an exterior door unit.
To replace the lock assembly with like kind.
To replace the glass panes in wood windows and doors.

OPENING COMMENTS

Conditions or circumstances that require corrective actions. For years the windows and doors in a home function as expected. However, they age and their mechanisms do eventually fail. In this chapter we will examine a variety of causes for these failures and an equal number of cures. Our concern is to retain the architectural integrity of the house while eliminating the problems.

The most often stated problem with doors and windows is their failure to operate properly. The conditions for these problems vary considerably, and we need to identify these for both windows and doors.

For example, the double-hung wood window shown in Figure 5-1 permits both the upper and lower window sashes to be moved up and down. (In a single-hung window, only the bottom window sash moves up and down.) Both window sash must be able to stay in position. When the owner opens the lower sash 3 to 4 in., it must stay there and not fall down of its own accord. Likewise, the upper window must stay up at all times, even when unlocked, and must also stay in any other position when moved.

Over the many years that these windows have been used (in this as well as other countries in the Western world), various methods and techniques have been used to hold the sash in place. The oldest and most common technique involved two pairs of lead weights. As the cutaway in Figure 5-2 shows, a pocket was built into the opening for the window unit. Two weights were housed in each side pocket. One weight was tied with either chain or clothes line; the end of the line was fed through a pulley and

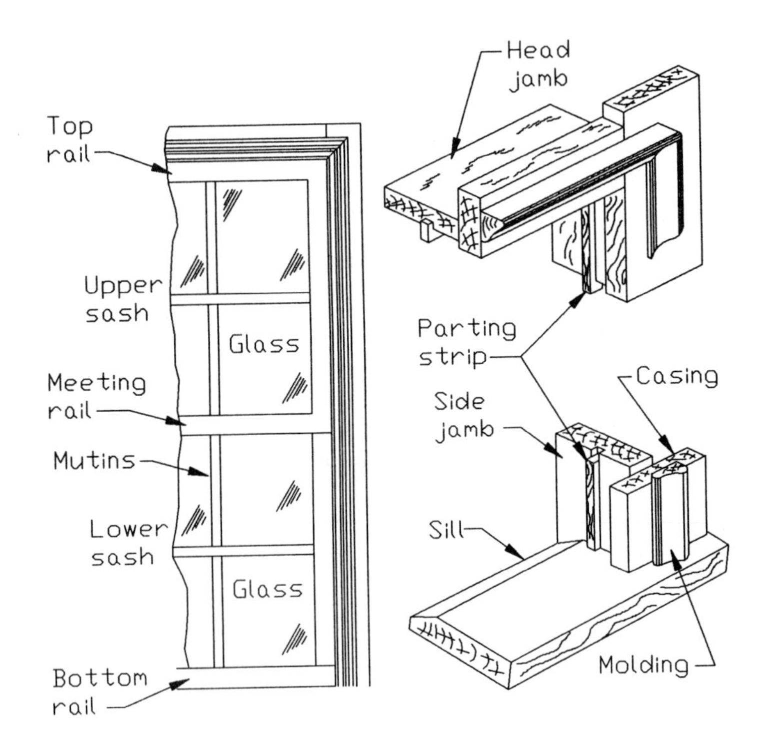

Figure 5-1 Double-hung window unit.

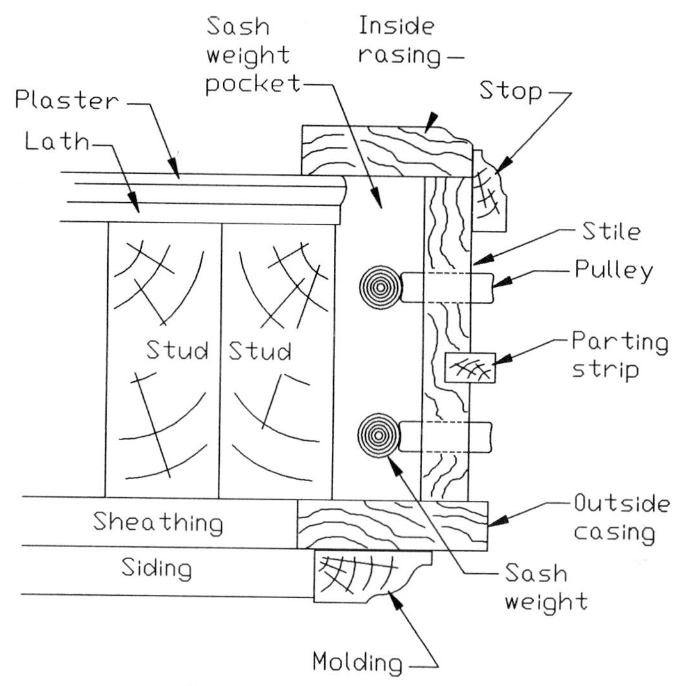

Figure 5-2 Pocket for sash weights

the free end was knotted and placed into a groove and socket in the side of the sash. The heavier weight was used for the lower sash, and the lighter one was used for the upper sash.

Over the years, the ropes were painted and decayed, and then they broke. The weight fell to the bottom of the well and the sash did not stay in place properly. If the upper sash rope broke on the left side, the window sagged toward the left when not locked in place, since there must be a clearance for the sash to travel up and down in its track. If the lower sash weight broke free on the left side, the sash would also sag to the left when the lower sash was raised. Of course, the opposite was also true. Thus it was always difficult to open and close the window's sash. The sash also looks ugly when they are canted. The only cure was to replace the ropes. This was no easy task, because the window had to be partially dismantled. Some homeowners mastered the task without destroying the parting strips and stops. Many ruined these pieces.

More recently, from 1950 to 1970, manufacturer's for double- and single-hung windows employed a cylinder assembly that operates from the coiling action of a spring and shaft. The cylinder is fastened to the window frame in the slot where the sash moves up and down. The top of the cylinder is nailed at the top. The moving part with its connecting arm is nailed under the sash. As the window is raised, the spring's tension holds the sash in place. In these windows, the common problem is that the nails work themselves loose under the sash rail and fall out. Then the shaft uncoils its approximate 10 to 13 twists and flies up into the cylinder. The result was that the window canted when positioned other than in a locked position. The cure was to reinstall the loosened shaft.

This type of control mechanism could also develop broken springs or spring holders. When this happened, the only cure was to replace the assembly. This meant taking off the stops and sometimes removing one of the parting strips.

Still more recently, the manufacturer's of window units replaced the cylinder with pressure strips of metal. These kept the sash in place with equal pressure on both sides. Friction held the sash. However, over time and especially with paint on the metal strips, the tension decreased and the window sash refused to stay in place. These windows were not easily repaired by homeowners since the replacement tension strips were not readily available.

The latest industry standards employ the tension principle but with a modification. The metal wraps entirely around the sash rail, thus increasing the total area of contact. It is difficult to paint this metal, and it should not be painted. Rather, it should be lubricated occasionally. Lubrication blocks air and retards much opportunity for rusting and oxidation. Windows with this new technology are also much more airtight.

Besides the double- and single-hung window units, sliding types also have their own problems. Sliding windows require grooves and insulation to prevent leaks of water and exposure to humid air. The tracks must have weep holes on the outside to allow water to quickly and completely drain away. But as friction increases because of wear in the track, the sash requires more energy to move it. This causes the owner considerable frustration, because there is almost always a kitchen cabinet or bathtub or commode to reach over. Painting the window also contributes to the problems.

The awning and jalousie types share a common problem after many years of use and wear. The cranking mechanism either breaks or the shaft holding the handle strips and no longer permits the crank to operate. These windows also have the common problem of not sealing tightly since there is only a crank on one side and it is this side that fits snug, while the opposite side does not. These windows are not easily repaired.

Aluminum windows are used extensively in home construction. When the first ones were made and installed, they were predominantly single-hung types with single-thickness glass and had snap-in mutins to simulate the small-paned glass of wood window sash. These windows employed a modified spring mechanism to permit the lower sash to be opened and remain so. Many used a cable attached to the bottom of the sash. It passed through the frame to a spring take-up assembly. When it broke, it was very difficult to replace. If the spring failed, the job was worse. Most of the glass was installed with putty or glazing compound. Some, however, was held in place with caulking compound.

Finally, all early window panes were installed with putty and had glazing points installed before the putty was applied. When the putty dries and cracks, it fails to hold the glass in tightly. The glass rattles and leaks air. Water also gets behind the putty and rots the wood frame of the sash. If the putty is replaced before the water causes rot, a homeowner can easily perform the job. When the rot is extensive, the sash must be replaced. If a replacement sash is available, the task of replacement is difficult but not impossible for the owner. However, if the sash is not readily available, then a wood or cabinet shop must fabricate one. The owner can then use the expertise of a carpenter to install it or can try to do it himself or herself.

Doors and door units have other problems. Exterior doors must fit well and must be well insulated. All too often doors warp, expand with wet weather, and contract with prolonged periods of dry weather. Wood doors today do not as a rule stand up to the direct onslaught of weather. For example, the sun's drawing power plays havoc with paint, varnish, stain, and just about any finish. Thresholds and saddles come loose or wear badly over time. Weatherstripping also fails to perform its function due to abuse and painting. Hinges become loose, locks fail to work, and striking catches work loose. On plywood doors, the outer veneer peals away. Glass panes come loose from the molding and putty holding them in place. The list is long. But for every problem there is a cure. Some of them can be made very well by the owner, but others need the carpenter's skill and knowledge to make it right again.

Both the door unit and window unit are built today much as they were years ago, but are much improved over the earlier designs. All styles remain much the same except for the jalousie, which has been phased out. Therefore, an alternative unit is an option for the homeowner. Instead of making extensive repairs to old units, new, improved ones with the same architectural style can be substituted.

Contractor responsibility. The contractor has the responsibility of informing the owner of a problem with a window or door unit by providing a clear explanation of the effort of repair, cause for the problem, and any problems there might be with obtaining replacement parts. With the information about windows and doors given above, the contractor can adequately solve the problem for the owner.

However, the contractor may have difficulty when replacement parts or sash need to be obtained. Modern manufacturers have changed their lines of products in favor of the conservation demands of today's homeowner. Many local companies make their own aluminum window units, for example, with pieces supplied by the aluminum companies. These assemblers might be able to rebuild a damaged window or find suitable substitutes for them.

When the problem is the entrance or back door, the contractor may need to special order a door to match the design of the original. This may be the only way to solve the problem. The owner needs to know that there will be a significant delay if a door must be built. Also, the locks may not be exact replacements if not manufactured anymore.

Homeowner's expectations. The homeowner usually does not need much explanation of the problem with the windows and doors. They don't work properly and need to be fixed. Sometimes the contractor is called after the owner or a member of the family has attempted repairs. When this happens, the problems are usually much worse.

The owner expects quick restoration and a fair price. But he or she must understand that the job is usually labor intensive, since carpentry skills must solve the problems to make the window and or door function properly.

Scope or types of projects to solve. In this chapter we will solve problems with damaged wooden windows, aluminum windows, and exterior doors. Each project has several problems associated with the window or door. This provides the reader with solutions that are adaptable to a variety of conditions.

PROJECT 1. RESTORATION OF A WOODEN DOUBLE-HUNG WINDOW UNIT

Subcategories include replacing sash cords or chains; freeing paint-shut sash; reglazing loose glass panes; replacing the bronze strips or other insulation; replacing cylinder sash controls; replacing awning controls; refitting window stops; stripping sash of paint; replacement of sills and aprons; replacement of stops.

Primary Discussion with the Owner

Problem facing the owner. The owner of the house has been faced with the problem of broken sash cords for some time. In some of the windows, one cord is broken, and in others, two are broken. Some of the upper ones have been broken for years and other sashes have been nailed and painted shut. Other lower sashes have been propped open to keep them from falling. They rattle and are drafty in cold weather. If it were not for storm windows, in winter the draft would blow through the house and greatly increase fuel costs. In making his or her appraisal, the contractor determines that every window needs work to restore them to sound condition.

Alternatives to the problem's solution. The alternatives are two; first, restore the present sashes by refurbishing them and the frames as well. Second, replace the windows with modern double-pane wood or vinyl-clad wood windows.

Different costs are associated with each alternative. The restoration effort involves the removal of each sash, reputtying the glass, cleaning off the buildup of paint, replacing the cords, and reinstalling the parting strips, sash, and stops. Then a fresh painting system of a primer and two top coats must be applied. The effort when replacing the windows with modern ones involves the removal of the old windows, which is no simple task, and the modification of the opening to receive the new windows, which must be a close approximation of the original window for size.

The fact that the owner has storm windows satisfies the problem for the winter in that the R-value is improved over having only a single thickness of glass. With the window repaired and fitted snugly, a dead air space exists between the storm window and window. This acts as an insulation. When the home is cooled in the summer, the owner can elect to leave the storm windows in place and, by doing so, reduce energy costs.

In the second solution, much more time and effort are needed to remove and reinstall the windows, which adds to the overall cost. Then we must add the price of the modern, well-insulated window. These are expensive even by today's standards. If the window has double-pane glass, the R-value is better than the single glass and storm window together. If the window unit has triple glass, the R-value is even better. Various manufacturers place the payback period for double-pane glass at between 6 to 10 years. What this means is that the energy losses through the window are calculated on an average basis per year; then the average utility or heating and air conditioning costs are multiplied against these losses. The total dollar value is then matched against the cost of the modern window and its installation. Thus, for example, if the annual savings are calculated at $25.00 and the new window and installation cost $200.00, the payback is 8 years. After 8 years, the owner begins to realize a savings. Some advertisements take into account the price creep for energy, and when this is factored in, the payback time is less, since the windows were paid for with cheaper dollars.

This leaves the owner with a dilemma as to which solution to select. Where there is great concern about the cost of repairs or replacement, the repairs will almost always be cheaper. This could be the deciding factor. If the present owner does not expect to live in the house or own it for a period of time sufficient to either break even or realize a savings, the option once again is to make repairs. On the other hand, if money is not an object or if there is too much wear and decay, the solution to replace the windows would be the best.

For our generic project, we will establish a contract for the repair of the old sash and frames and repainting them. However, in the section on reconstruction, we will examine the work effort and problems associated with removing and replacing a window unit.

Statement of work and the planning effort. The amount of effort associated with the restoration of each double-hung window is the same. Therefore if we can agree on the statement of work for one, we can apply the cost to each window in the house. So we will *restore each double-hung window in the house to a sound, operable condition.* This general statement is further broken into specifics such as *all sash and frames will be stripped of paint and later repainted with a three-coat system.* Another statement of work adds the details that *all cords will be replaced, and when the sash are reinstalled, the stops will be installed snugly.* We might even add that *locks and handles will be replaced or cleaned and made operable.* Finally, we can add that *no residue of paint will be left on the glass, whether old or new, and any slightly damaged or rotted wood will be repaired with a wood filler appropriate for exterior exposure.*

The planning effort will not require a great deal of estimating, but some must be done. Obtaining supplies may not be that difficult, since the work is primarily restoration of the present windows. We will, however, need to locate a carpenter who has experience with working on these types of windows. The windows are old and their materials are subject to easy breakage. In fact, we will plan on replacing the parting strips and stops as a cost-saving effort.

Once the window sash have been broken loose from the frame, the work can be done inside the house. Therefore, scheduling must be done with the owner. Someone must be home.

Stripping the paint and reglazing the panes need to be done in the garage, under the carport, or in the driveway. For this we need a workbench, chemicals, protective gloves, and proper ventilation. We need to make repairs to the sash where wood has rotted away while the sash are out of the frames.

While the sash are out, we must replace the cords, lubricate the pulleys, and clean the frames.

Then we must make arrangements for the painter so that the painting system can be applied to each window. Where the carpenter will take an average of four hours to free the sash, make the repairs to the windows, and strip the paint, the painter will require three trips, one to prime the windows and two to apply the first and second top coats of latex.

Contract. The contract between the owner and contractor includes the price for the painter as well. This is such a straightforward job that a simple fixed-price contract will suffice. The wording in the body of the contract should include the number of windows to be refurbished as well as a description of the work. It could be as simple as this:

> For the price stipulated below, we agree to refurbish all double-hung windows in the house. This work includes:
>
> 1. Replacement of window sash cords.
> 2. Replacement of wooden parts that break, such as parting strips and stops.
> 3. Removal of built-up paint and a three-coat repaint.
> 4. Replacement of putty on all panes that require it.
> 5. Reinstallation of window locks.
> 6. Refitting sash to minimize air leaks.

Material Assessment

Direct Materials	Uses/purposes
Sash cord	Replace the old and broken sash cord
Paint remover	Strip paint
Latex primer	Prime the sash and frames
Latex exterior	Top coats
Parting strips	Replace those that split or crack
Window stops	$1\frac{1}{8}$-in. Stops to hold the lower sash in place
Glazing putty	Reputty the sash panes
Glazing points	Replace rusted and missing points
4d Finish nails	Nail parting strips and stops in place
Sash locks	Replace damaged and missing locks

Indirect materials	Uses/purposes
Mineral spirits	Clean up for some strippers
Rags	Clean up
Sandpaper	Smooth surfaces after stripping and between coats of paint
Mason line	Used as a feeder to snake the cord through the pulley
Oil or penetrating oil	Free and oil the pulleys

Support materials	Uses/purposes
Workbench	Ease the work of puttying, making repairs, and stripping

Support materials	Uses/purposes
Step ladders	Used to free the sash on the outside and paint the window
Extension ladder	Used to free the sash on the outside and paint the window
Carpentry tools	Construction
Painting tools	Painting

Outside contractor support	Uses/purposes
Painting contractor or painter	Apply the painting system

Activities Planning Chart

Activities	Time line (days)							
	1	2	3	4	5	6	7	8
1. Contract preparation	X							
2. Scheduling and materials	X							
3. Removing the sash		X	X	X	X	X		
4. Refurbishing the sash		X	X	X	X	X		
5. Refurbishing the frames		X	X	X	X	X		
6. Reinstalling the sash		X	X	X	X	X		
7. Applying the paint system						X	X	X

Reconstruction

Contractor support. Once the contractor has made the inspection and clarified the work to be done, he or she needs to take a small amount of time to prepare the contracts and work with the painter in establishing the price for painting. This job is labor intensive, so labor will be a large part of the bid/contract price. Materials will be smaller by comparison. Even though the overall price of the job will be relatively modest, allowance for fixed and variable overhead must be applied and an allowance for profit should be added.

Materials and scheduling. The carpenters and painter can pick up the materials required for the job on the first day of work or the night before since there are only a limited number of items. All the items are readily available from builder's supply stores or lumberyards. The painter may get his or her supplies from a paint specialty or builder's supply store.

Notice in the activity list that tasks 3 through 5 are parallel. They all start on the first day after contract agreement and continue for five consecutive days. (Note that we are predicting that the carpenter will restore 2 window units per day. If there were more than 10 window units in the owner's house, the time line would be extended. Also, if two carpenters were sent to the job, the time would be one-half or maybe less for 10 windows, since some economies of scale can take place.

Removing the sash. Sashes that have broken cords and/or are painted shut must be freed with care to avoid costly damage to the window frame and sash, too. So, let's start by removing the lower sash first.

Removing the Lower Sash. The carpenter begins work on the window unit (we will refer to the window unit as a window) as follows:

1. Remove the curtains, blinds and their hangers, if any, and window shades and their hangers.

2. Place the claws of an 8- or 12-oz. hammer behind the stop and pry each loose. We usually start just above the lower sash. There is a strip of stop on each side. This piece holds the lower sash in its track.

3. With the stops off, we can lift the lower sash out of the track. If it is painted shut, we will need to go outside and use a flat bar or wood chisel to carefully crack it free. Most of the time the job is made safer if we first take a utility knife with a new blade and cut through the paint in the corner between the sash and frame.

4. The sash only comes out so far because the ropes are holding it. As Figure 5-3 shows, the rope is knotted and held in place with a nail. The cord fits into the track grooved in the sash style. We usually cut the cords with a pair of lineman's pliers, but we could also pry the nail free with a common point screwdriver and pull the cord free. *Caution:* Since there is a 6-, 7-, or 8-lb. weight on the other end of the cord, we must not let go. We must ease the weight up to the pulley. With the cords free, the sash is out and can be taken to the workbench for further work.

Removing the Upper Sash. The carpenter will free the sash first and lower it before taking any steps to remove the sash from the frame.

1. We must remove a parting strip to remove the upper sash. To do this, we insert a sharp chisel (3/8 in.) between the strip and window frame and pry it free from its groove. There is only one 4d finish nail in this strip and it is usually about midway up the strip. Since we are planning to replace these strips, we do not need to be extremely careful about breaking them. However, we should avoid marring the frame.

2. With the parting strip out, we move the sash out until we can reach the cords. We remove these as discussed before for the lower sash.

3. We then take the sash to the workbench for further work.

Refurbishing the sash. Generally, as carpenters, we perform the following tasks when refurbishing the sash. We clean off the old buildup of paint to determine if there is rotted wood that either needs to be cleaned and filled or replaced and also

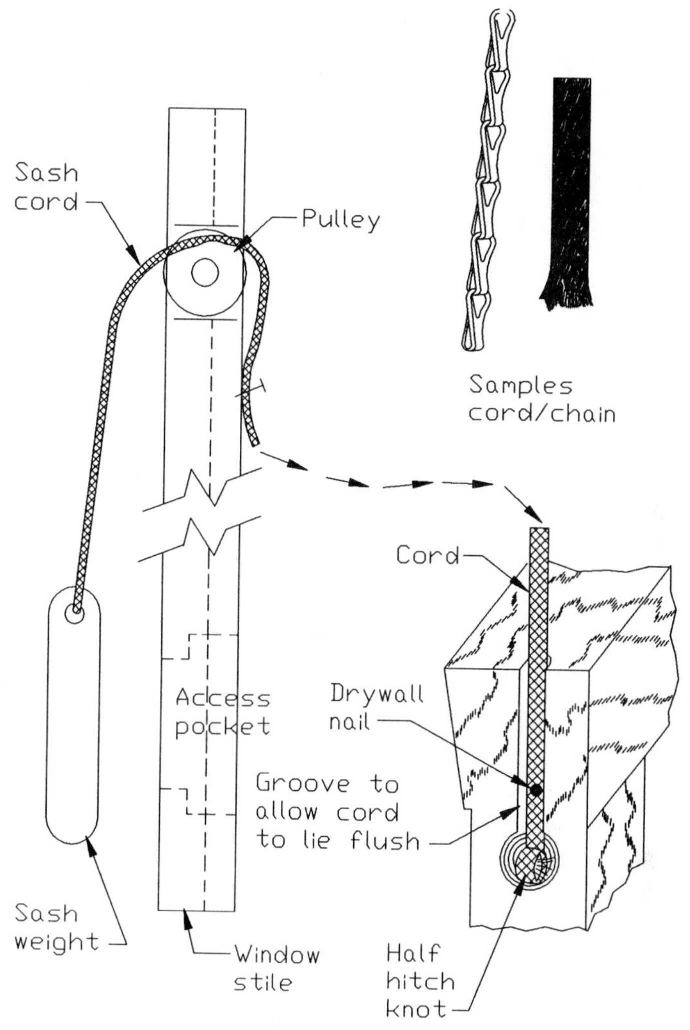

Figure 5-3 Replacing ropes or cords in a sash.

to make it easier to apply new putty. We also remove the old ropes from the sockets and the lock from the top of the sash.

Then we clean out the old putty, ensuring that the glazing points firmly hold the glass in place; if not, we use new ones to do the job. We then apply new putty with either a putty knife or 1-in.-wide chisel. The putty must meet the specifications that one edge is flush with the wood as in Figure 5-4a. The putty touching the glass must be even with the wood's edge, as seen by looking through the glass (Figure 5-4b).

Refurbishing the frames. An inspection of the frames must take into account the condition of the window sill and all the frame pieces. We will check for split or decayed pieces; to do this, we need to remove most of the paint. Then we remove

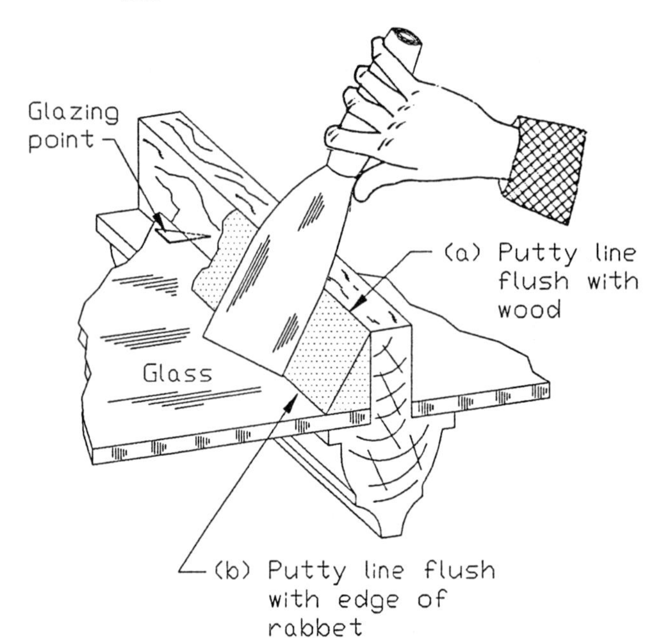

Figure 5-4 Applying putty to a sash (a) flush with wood edge, (b) even with edge of window glass.

the other parting strip to reach the wood piece that lets us get to the sash weights. Figure 5-5 illustrates this. There is one screw (usually a 3/4-in. No 9 flat head) holding the piece in place. We usually remove the screw and pry the piece out. Then we reach inside and find the two sash weights resting on the bottom sill frame. The heavier one is used for the lower sash.

We inspect the pulleys and lubricate them. If they are frozen with rust, we use a rust remover lubricant and work the pulleys loose. Then we use a length of mason cord with a weight such as a 10d or 16d finish nail as a feeder to string the new sash cord through the pulley and into the sash weight pocket. Figure 5-6 shows this arrangement. Once we can see the nail and line, we pull it out of the pocket and tie the sash weight to the end. Next we pull the cord and raise the pulley to the top. Then we drive a 4d finish nail through the cord and partially into the frame. We then cut the end about halfway down for the upper sash and at the sill for the bottom sash. Once each sash weight has a new cord, we reinstall the access piece and secure it with the screw.

If the window sill or any other piece of the frame is rotted, we must replace it. Since these pieces were made many years ago, when dimensions of lumber were thicker, we will need to custom cut and fit each piece. This requires skilled workmanship.

Reinstalling the sash and trim. With the frame repaired, cords installed through the pulleys, and sash retrofitted, we are ready to conclude the carpentry work. The steps are as follows:

1. Install the upper sash by trimming the cord and making a knot in each piece. Press the knot into the hole made for it, and either make sure that it is secure in the

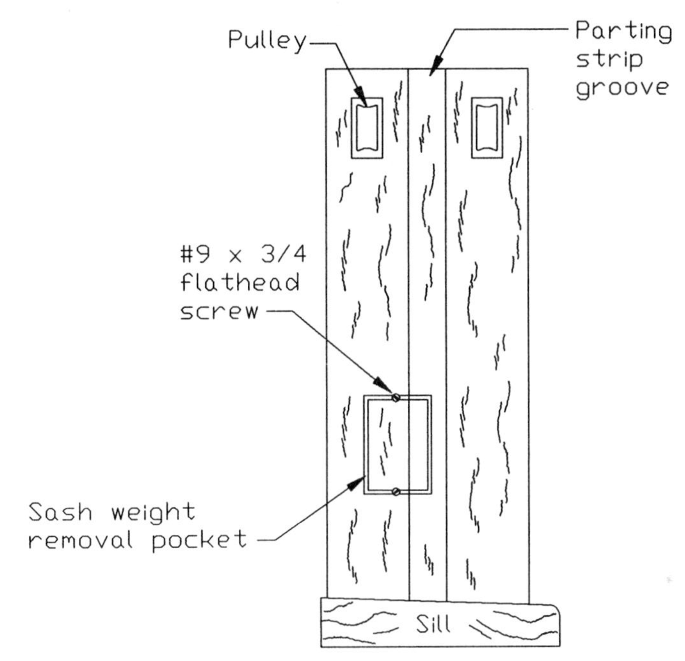

Figure 5-5 Access port to sash weight pockets.

sash cord fixture or drive a drywall nail through the cord into the frame just above the hole. Position the sash and lower it as close to the bottom as possible. Install the parting strips. Raise and lower the sash to check for even, smooth operation.

2. Install the lower sash by installing the cords in the pockets on each side of it. Position the sash in the down position. Cut and fit a new stop for each side. Tack the stops in place and test the sash for free but not sloppy up–down travel. Move the stops as required. Drive the nails home and set the heads.

3. Install the hardware. Reinstall the lock and handles if there are any.

Painting system. The painter will take a minimum of 3 days to paint the windows. On the first day he or she will apply a prime coat of latex. On the next two succeeding days, the top and final coats of latex will be applied. The colors must match the interior woodwork. The colors on the outside must match the trim of the house.

Prior to applying any paint, the painter will fill holes and small cracks with putty or spackling. This product needs to surface dry before it is painted. Latex paints are cleaned up with water and detergent.

Concluding comments. Restoring a double- or single-hung wood window is a time-consuming job. Much care must be taken to avoid damage to both the sash and frame. Using a workbench makes the work easier and efficient.

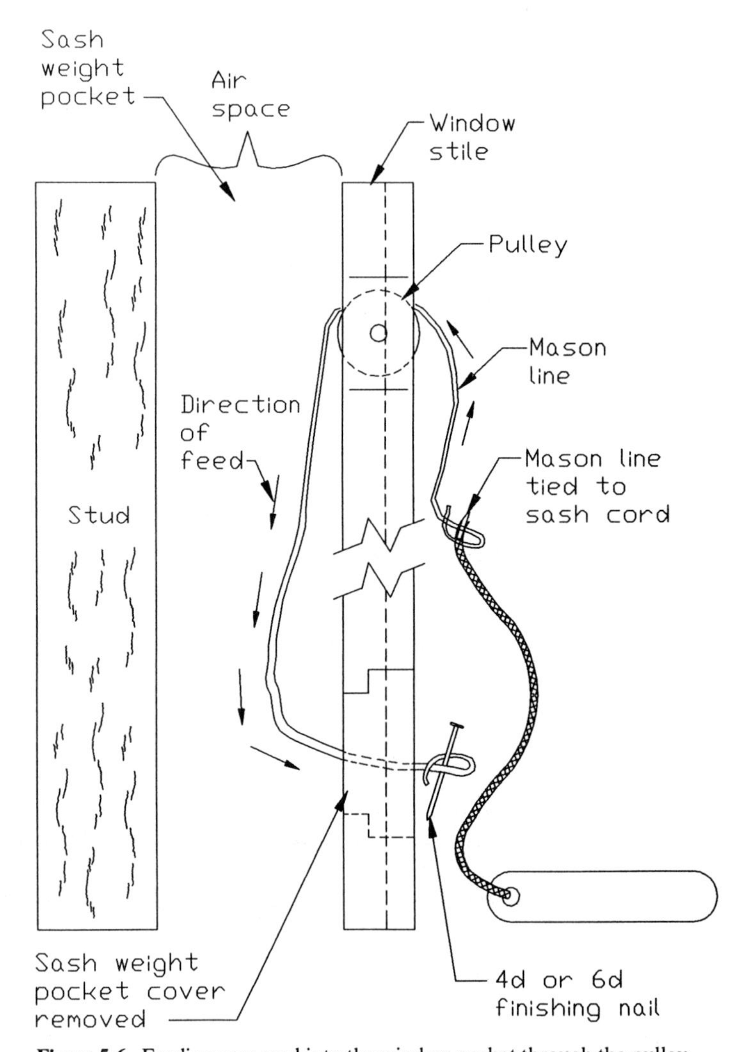

Figure 5-6 Feeding new cord into the window pocket through the pulley.

The best job is done when the old paint applied to the window is stripped to the wood first. This reference is essential to reglazing, locating rotted parts, making repairs, and refitting the sash to the frames.

Replacing the Window Unit

Removing the old window. To replace the window unit, we must remove the frame as well as the sash. Usually we can use the same set of tasks for both wood and aluminum or metal window units. The exception is those set in mortar.

The tasks require that we first remove the casing, apron, and window sill from the inside. If the window unit has weights, the cords must be cut and the weights allowed to drop into the base of the pocket or removed from the pocket. Then, working from the outside, we try to pry the frame loose from under the outside sill with a flat bar or crowbar. Many times this technique works well. In some situations, we can drive the nails through the trim most of the way, and this makes the window easier to pry out.

When the outside casing is behind the siding, we are faced with a more difficult task. We need to either dismantle the frame, cut through the casing after removing the molding, or pry out the casing after freeing it from the style of the window frame by driving the nails through with a nail set. Even this technique may not entirely free the window, since the siding may also be nailed into the casing.

To dismantle the frame, we first cut the sill and head in two. Then we can remove these parts. (This assumes the sash are already removed.) Then we can pry the stiles out, and we are left with the outside casing and molding. Close examination will provide the answer about removing the casing or trimming it flush with the siding or molding.

Steel and aluminum windows have a small strip that extends out on both sides and top; it is used for nailing and making a watertight corner into which the siding can fit. When these windows are removed, this metal strip bends readily and tears as the prying operation takes place.

Installing the new window unit. Installing the new window unit requires some decision making on the site. Problems of matching the opening must be resolved, as must the problem of securely nailing the window in place. Then, too, we must make the window watertight with proper installation and some form of caulking.

After the new unit is installed, we are faced with more problems on the inside. On a rare occasion, we may be lucky and the old sill, apron, and casing will fit perfectly against the window unit. Most of the time, however, we must reposition these pieces or add fillers. This means that some touch-up painting is required.

Some owners have sufficient skill to do this work. Those who would try could use the techniques described here. The only difficult part is determining how to make the new window fit in the opening.

PROJECT 2. REPLACING THE WEATHERSTRIPPING ON WINDOWS

Subcategories include replacing the metal weatherstripping strips; replacing the composition weatherstripping pieces; refitting the stops to take slack out of the sash travel; adding new types of weatherstrip; using filler strips to reduce drafts; loss of heating and air conditioning.

Primary Discussion with the Owner

Problem facing the owner. Wood, vinyl-clad wood, and many varieties of metal windows are insulated to reduce heat and air-conditioning costs. If the owner has had the utility company make an assessment about the quality of window needed to stop infiltration of air, pollutants, and penetration of rain and dust, we need to either apply weatherstrip or replace the existing weatherstrip.

Weatherstrip has been used to seal windows from the earliest part of the century to the present. In the 1930s, for example, zinc-clad metal was installed on the window sash and frames by carpenters and weatherstrippers. Today, all reputable window manufacturers build their windows with insulation that performs the same function as the carpenters did manually in the 1930s.

Manufacturers have various types of insulation materials made to their specifications. A few of these are spring-tension pieces and sash guides, where the tension forces the metal away from the frame and against the sash, and woven felt, where the felt is glued to a metal backing and the seal is made when the felt is compressed. Another is a compressible bulb that compresses when the sash is lowered.

The contractor needs to examine each window to determine the type of weatherstripping used. Then he or she would need to locate replacement materials. The owner needs to know that this may not be a simple task, since there are many window manufacturers besides the more well-established and better known ones. There are few standards for weatherstripping except those established by each company; it must only meet its intended purpose.

The contractor needs to show the owner where the problem with leakage occurs, which might involve any of the following:

1. There never was any weatherstripping.

2. The weatherstripping has been painted repeatedly and thus cannot spring; the tension is gone.

3. The compression is gone in the woven felt.

4. The woven felt has abraded away.

5. The softness in the compressible bulb has lost its elasticity and/or was painted and is too stiff to compress.

6. The wraparound compression strip has been damaged and does not make full contact with the sash or frame.

Alternatives to the problem's solution. The owner must try to have the exact replacement weatherstripping reinstalled in each window. But, when the exact replacement cannot be found, alternatives need to be tried. These alternatives must provide the same degree of protection as the original materials did.

The contractor may be able to obtain a replacement strip whose characteristics are very close but the size is off just a bit. In this case, careful trimming may result in a satisfactory job. If spring tension strips were originally used, a metal shop or metals manufacturer may be able to custom make enough to retrofit every window in the house. The contractor would likely need to send a sample of the old metal to the manufacturer. A third possibility is to use an alternate system from a window manufacturer. This may be learned by contacting the manufacturer's retail outlet or district representative. It is possible that the replacement will work without modifying the sash or frames. On the other hand, a combination of using a different system and making adjustments to the sash or frames will produce a very satisfactory job for the owner.

Can the owner perform this work? Well, many can who have considerable skill and the time and energy to search for replacement materials. If the owner has the time off from earning a living, then the solution is "do it yourself." If, however, it costs more in salary to locate the materials and perform the restoration, then it is economically wrong to do it yourself.

Statement of work and the planning effort. The statement of work for this effort must include *replace the defective metal and insulating materials in each window in the house.* This single statement may be sufficient. But other aspects may need clarification. One of these is *repaint any materials removed to make access to the sash and installation materials.*

The planning effort must include assigning someone in the office to canvass the local builder's supply or window manufacturer's outlets in the region or to make a call to the parent company if the maker of the windows is known. If the exact metal or other type of insulation cannot be located, either a separate contract must be made with a specialty shop or an alternative solution must be defined and estimated. These several aspects will add to the overall cost of the restoration effort. The owner must expect to pay for the search.

Once the materials are located and their prices are known, the contract can be submitted. Along with contract submission, we advise the owner that someone must be home to permit the workers access to the house while the work is in progress. We also provide a schedule of the expected number of days needed to complete the job.

Contract. This contract would be a fixed-price type if any of the following conditions prevailed:

1. Materials are readily available from a known source.

2. Suitable alternative materials are located from a known source and the owner has agreed to their use.

3. Materials have to be locally manufactured and the shop owner can provide a fixed price for their manufacture.

These conditions would have to be agreed to before the contract was originated. This is fair to both parties in arriving at a reasonable cost.

If none of the three conditions prevails, then the time and materials contract will be more appropriate. The contractor will undertake the job and document the activities used in locating the materials. He or she will also charge an hourly rate to restore the weather-soundness of each window. The painter will also work by the hour.

The contractor must estimate the time to restore each window based on similar work done before or by the use of a standard that might be available from a specialized source. A good estimate to restore a double-hung window would be an average of 4 hours. A single-hung or awning-type window would require less time. The first one or two will take somewhat longer, but once the technique is mastered, the work will go more quickly.

Material Assessment

Direct materials	Uses/purposes
Weatherstripping	Replace the worn and defective materials
Nails	Hold the new weatherstrip in place
Caulking	Seal cracks around window frames, as required
Paint	Repaint the stops and trim
Spackling	Fill nailhead holes before painting

Indirect materials	Uses/purposes
Window stops	Replace cracked or broken ones
Parting strips	Replace broken ones
Putty	Reputtying the glass

Support materials	Uses/purposes
Carpenter tools	Construction
Ladders	Access the upper parts of the windows
Drop cloths	Protect the floors

Outside contractor support	Uses/purposes
Painter	Paint and touch up
Metal shop (optional)	Make the metal weatherstripping

Activities Planning Chart

Activities	Time line (days)						
	1	2	3	4	5	6	7
1. Contract preparation	X	X					
2. Scheduling		X	X				
3. Restoration			X	X	X	X	
4. Repainting and touch up						X	X

Reconstruction

Contract preparation and scheduling. We have already noted some of the preparation activities associated with contract preparation. Before the actual work begins of preparing the document, time is used for locating the materials. Work of this nature must be added to the overhead costs, although the time could be included with direct labor. Local telephoning is an indirect cost of doing business and is apportioned according to the schedule for the year or by the job. If long-distance calls need to be assigned to this job, they would be a single line item in the shop estimate. They would not show up in the final fixed-price contract as an item.

The work schedule for the restoration should be in hand at the time of the contract signing. The owner needs to agree to the schedule since someone must be home to allow the workers access. From our action plan, we see that a schedule of 4 days is required for the restoration. We already stipulated that for a double-hung window an average of 4 hours is required to restore the window's quality weatherproofing. Thus one worker is required if the house has 8 windows, two workers if it has 16, and so on.

The painter can begin work immediately after the carpenter is finished with the window. However, this would be wasteful and expensive. Therefore, the painter is scheduled to arrive on the last 2 days of the contract. The first day he or she will prime the work and do touch ups. The last day the final coat or coats will be applied.

Restoration. In completing the restoration of each window, the carpenter must remove the defective parts of the window. This will probably include removing the sash and some of the trim. Stops and parting strips need to be taken off with extreme care to avoid breaking them. Sash cords or spiral spring tension units must be carefully removed. The old stripping and metal pieces need to be removed and set aside.

There are usually a variety of things in the window tracks, such as dirt, debris, overrun of paint, broken metal, pieces of insulation and weatherstripping, and the like. We remove these particles with a stiff wire brush and blow the channels clean before starting to install the new materials.

In a double-hung window, we always start with the upper sash. The three pieces that go into the frame are the head piece and two style or side pieces. Before reinstalling the upper sash, we must replace the weatherstrip on the front or inside of the meeting rail. Then we reinstall the ropes or spiral controls and test the sash for proper operation.

Then we install the metal into the lower sash track: first, the bottom that sets on the sill and then the two side pieces. These are all held in place with one or two 1-in. flat head nails provided by the vendor. Before we set the lower sash in place, we fasten the matching weatherstrip to the outside of the meeting rail. Then we reinstall the sash cords or spiral springs and test the operation.

Finally, we nail the stops in place. These should fit snugly against the sash but should not prevent it from sliding up and down.

Concluding comments. The replacement of damaged or worn-out weatherstripping is a task that requires a considerable amount of know-how on the part of the carpenter or layperson. Since no additional damage must be done to the window frame and the sash, no part of the job can be done hurriedly.

Obtaining the necessary materials can be almost as difficult as performing the work itself. The older the window, the more likely a substitute weatherstripping will have to suffice. When this is the case, we find that other modifications such as padding might be used. Also, while finding a source for the weatherstripping, we should order an additional 15% to 20% to allow for mistakes while doing the installation and to preclude having to delay finishing the job by having to make a second order.

Although we used the double- and single-hung window as an example in the study, the process works for every type of window, including aluminum and vinyl clad, too.

PROJECT 3. RESTORE OR REPLACE A DAMAGED EXTERIOR DOOR UNIT

Subcategories include decayed exterior door trim, worn or rotted threshold or sill; split or cracked jamb; defective or missing striking plate; stripped screw holes in either the door or frame; broken or inoperable locks; damaged door surface; warped door; ill-fitting door; broken panes of glass; damaged or destroyed weatherstripping; broken door stops; separated stiles and rails. Figure 5-7 shows an exterior door unit with the parts identified.

Primary Discussion with the Owner

Problem facing the owner. From the list of subcategories, we see that the owner can be faced with a multitude of problems with exterior door units. These problems can exist on standard 36-in. front door units, standard 32-in. side or rear door units, and 24- to 34-in French door units. Where double doors are in a unit, added problems can and do happen.

The owner of the home usually allows problems with the door unit (door and frame) to become severe before calling for help. The changes from a snugly fit, solid, and weatherproof unit into one that leaks air and dust, does not close except when slammed, or suffers other more serious problems happen over an extended period of time. The changes are sometimes so gradual that children grow into young adults before repairs are made.

When the contractor is called to the home to solve the problem, the owner is upset with the door because it won't work and they want it fixed "right now." The fact that the restoration could cost from as little as $75.00 to well over $1000.00 does not usually enter their minds. So we, as contractors, need to understand a little about human nature. We must carefully examine the problem door and frame. We need to first ask questions about the problem and how long it has been coming on and generally to establish that the problem has a certain amount of depth and longevity. This brings the owner closer to the extent of the problem and softens the alternatives as they are presented.

Alternatives to the problem's solution. Since so many things can go wrong with the door unit, let's make a list of problems and alternative solutions:

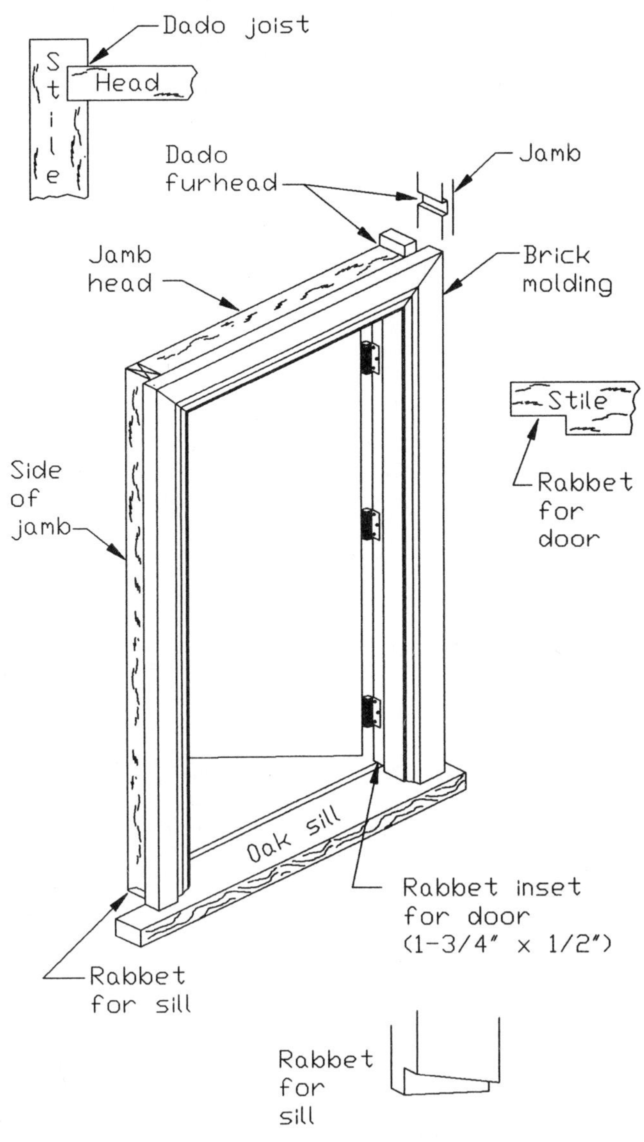

Figure 5-7 Exterior door unit.

Condition	Solutions
1. Decayed exterior door trim or jamb	a. Replace the defective trim b. Splice new wood after cutting away the damaged wood
2. Worn or rotted threshold	a. Remove and replace the wood or metal threshold
3. Worn sill	a. Remove and replace the sill b. Replace the rubber or metal seals

Condition	Solutions
4. Split or cracked jamb	a. Remove and replace the split stile or head b. Glue and screw the split jamb
5. Defective or missing striking plate	a. Replace the missing plate in the jamb b. Repair the striking plate area and reinstall the plate
6. Stripped screw holes in either the door or frame	a. Apply glue and a tapered dowel in each screw hole, reinstall the screw b. Replace the wood area around the hinge and reinstall the hinge
7. Broken or inoperable locks	a. Remove and replace the locks b. Disassemble and repair the locks
8. Damaged door surface	a. Install a patch in a solid panel door b. Replace the damaged stile or head with a custom-made piece c. Replace the wood panel (inset)
9. Warped door	a. Replace the door or frame b. Rehang the door to fit
10. Ill-fitting door	a. Readjust the position of the door by moving the hinges b. Spread or shim the jambs c. Add filler strips to the stops d. Add filler materials under the sill or threshold e. Trim the door to fit
11. Broken panes of glass	a. Remove and replace the glass
12. Damaged or destroyed weatherstripping	a. Remove and replace the weather-stripping b. Straighten out the bent metals c. Reseal the metal sill
13. Broken door stops	a. Replace the stops b. Splice the broken pieces
14. Separated stiles and rails	a. Separate the door, add glue and clamp, add screws or pins b. Replace the door

As we can see, there are many possible problems and many alternative solutions. Almost every one requires the skill and knowledge of an experienced carpenter, with plenty of experience with finishing work. Rough carpenters cannot, as a rule, perform the exacting detail needed to make repairs. New or apprentice carpenters seldom have experience repairing doors and their frames, since these people usually only install new premanufactured units. Finally, most homeowners lack the skill and knowledge. However, those that are handy can use this book and others, such as *Carpentry For Residential Construction,* Craftsman Book Co., 1987 by this author, to be guided through the repair.

Statement of work and the planning effort. In general terms, the statement of work should be *make repairs to the door and frame to restore it to its original condition*. However, this general statement of work may permit two meanings. First, the contractor would be able to take whatever measures necessary at any cost to restore the door unit. The owner, on the other hand, would expect minimal cost and minimal work to effect the restoration. Therefore, a more precise statement is better for both parties, such as *make repairs to the door frame and replace only the parts that are defective*. This sets boundaries. Likewise, the statement of work about the door could be *restore the door to its original soundness or replace it, whichever is least expensive*. Then a follow-on statement of work could address the fit of the door in the frame, such as *refit the door in the frame to operate flawlessly and also be weathertight*. These last three statements more clearly define the limits the owner is willing to contract for and also provide boundaries for the contractor. Under these limits, the contractor can perform a quality job.

Planning the work effort will not be difficult. However, the contractor will need to determine, within the limits, what the scope of work will be in terms of replacement materials and labor intensity. The door unit may be the kind used in developments and tract homes. This kind is usually of good to fair quality, but never ornate. (The contractor is always trying to reduce costs.) Since these types are readily available and their costs are moderate, the work in locating a replacement should not take more than several phone calls. If the contractor elects to refit the door and replace defective parts and weatherstrip, these materials should be readily available. Therefore, the contractor expects that a trip to the builder's supply store will secure all needed materials.

The planning effort in writing the contract should be relatively straightforward.

The schedule to do the work should only include 1 day for the replacement and not more than 2 days to restore the door and frame. Then repainting should not take more than two trips by the painter, but each day is less than a full day of work.

Contract. The contract can be a fixed-price type if the owner and contractor can agree on the *statements of work* that best suit the situation. The body of the contract should clearly specify that the door unit will be replaced or restored and under what conditions the restoration work will be done. Since painting is required, who will do the painting and who will supply the painting materials should be stated in writing.

The contract may or may not include a statement that requires the owner to have someone home while the workers are on site. It would be safer to make such a statement, but a mutual oral agreement can also suffice if there is a certain amount of trust.

Material Assessment

Direct materials	Uses/purposes
Alternative 1	*Replace the door unit*
Door unit	Replacement door and frame
Caulking	Sealant
Nails	Hold frame to wall
Casing (maybe)	If old casings are split

Stops	Replace old door stops
Threshold	Replace worn or defective threshold
Screws	Replace rusted screws
Hinges (maybe)	If frozen or rusted

Direct materials	**Uses/purposes**
Alternate 2	*Refurbish the old door unit*
Lock set	Replace if worn
Door casing (maybe)	If cracked
Door glass (maybe)	If broken or cracked
Glazing putty	Hold glass in place between glass and door routing
Panel (maybe)	If panel is cracked
Weatherstripping	Replace old, worn, or damaged
Painting system	Exterior latex painting system

Indirect materials	**Uses/purposes**
Shims	To move frame to align with edge of door
Wood plugs and glue	To fill old and worn screw holes
Mason drill bit	To drill new holes in concrete floor for sill

Support materials	**Uses/purposes**
Carpentry tools	Reconstruction
Power drill and bits	Reconstruction
Extension cords	Supply power
Workbench	Surface to make repairs to the door
Step ladder	Makes reaching simpler
Drop cloths	Protect the carpets and block cold or hot air entry while the door is off the hinges

Outside contractor support	**Uses/purposes**
Painting (optional)	Apply paint system to door unit

Activities Planning Chart

Activities	Time line (days) 1	2	3	4	5	6	7
1. Contractor office effort	×						
2. Remove and replace door unit		×					
3. Restore door and fit		×	×				
4. Apply paint system				×	×		

Reconstruction

Contractor office effort. The costs and effort associated with this project depend on the discussion with the owner at the time of examining the condition of the doors. If the contractor talked the owner into a replacement unit, then a smaller effort is needed to prepare the contract price-out. Several calls to supply houses will produce the best cost and availability. Another call to the painter, if the owner elects not to paint the door unit himself or herself, should produce a price for the job.

If, however, the owner wants the door restored to a quality condition, then more effort will be required to locate the necessary materials. We will need to locate all the materials listed in the materials assessment under alternative 2. The difference in effort will be translated to overhead and charged to either materials or labor or both.

Remove and replace the door unit. This is alternative 1. We pick up a new door unit at the builder's supply store or have one delivered to the job site. Along with the unit, we must have caulking and nails sent along. The caulking is used between the wall and backside of the trim to make a seal against water and moisture intrusion. The nails (usually 10d galvanized finish or casing type) will be used to nail the door unit in place. A threshold is also required and it is ordered with the door unit.

The work goes like this:

1. Remove the inside door casings.

2. Remove the threshold. It may be nailed or screwed. It may be wood or metal.

3. Use a block of 1/2 in. × 2 in. about 1 foot long and drive it between the jamb and stud frame. The idea is to drive the door trim away from the wall. If there is no room for the wedge, then use a claw hammer and block of wood and gently pry the door trim away from the wall.

4. After the trim is loosened on all three sides, pry the door unit all the way off and remove it from the opening.

5. Clean the area. Remove the nails that pulled through, sweep the threshold and sill area very clean, and remove any nailed blocks or spacers on the framing.

6. Lift the new unit in place and test for fit.

7. Determine the amount of adjustment needed to raise the unit in place to ensure that the door will clear the carpeting.

8. Check the depth of the unit to be sure that the jamb does not stick in the inside too far. It usually does not. But it frequently does not come flush with the drywall. Therefore, we will be required to add filler strips before installing the casings.

9. With the door centered or in the same place as the old one, we nail it in place with the 10d casing or finish nails without leaving hammer marks; use a nail set to drive the nail home. Before driving the nails home, we need to test the door for proper swing.

10. Next we cut blocks to fit between the jamb and framing. There must be one behind the hinges and behind the striking plate, as a minimum. Again, we test the door for fit (1/8-in. clearance) on each side and across the top.

11. If the fit is perfect, we nail through the jamb into the framing at the areas where the filler blocks are.

12. Next we install the threshold. This may have come with the door or, if not, we purchased one.

13. The threshold overlaps the sill of the door unit and the carpeting in the house. Most of it is over the sill.

14. The threshold is exactly centered under the door. The door's seal must come in contact with the threshold when the task is finished. Therefore, we may need to pad the underside of the threshold with tar paper. *Note:* This is a special consideration and almost every installation requires a special solution that carpenters trained for the work can do.

15. Next we install the door casings.

16. Finally, we install the lock set and test the door again for proper operation.

Restore door and frame. The restoration of a door unit may be more or less labor intensive. It depends on what must be corrected. So we need to examine each possibility separately.

Door Fails to Stay Closed and Locked. This is one of the most aggravating problems. Something has changed on the door unit to make this happen. We need to locate the problem and correct it. The four most usual problems are as follows:

1. The door has warped and, because the bottom leading edge hits the stops first, it takes excessive force to push the door closed far enough for the bolt latch to reach the striking plate. The cure is to readjust the hinges to lessen the warp on the leading edge. By moving the bottom hinge or bottom and center hinge out (toward the inside), the top moves toward the stops at the top. The door closes much easier and latches.

2. The striking plate has moved. This only happens if the plate was poorly installed. When properly installed, the plate cannot move since it is set into the wood and screwed in place. However, if the plate has moved, it can be reset accurately.

3. The bolt on the lock assembly jams because the lock unit has shifted or the spring is bad. In any event, the usual cure is to replace the lock set.

4. The weatherstripping around the door has moved and does not permit the door to close against the stops. The thickness of the weatherstripping is sufficient to prevent the lock from catching the striking plate. The solution is to replace the weatherstripping with new materials.

Door Is Not Weathertight. The solutions are few:

1. Replace the weatherstripping. Remove the old materials that have been painted and are worn out. Install new materials.

2. Replace the composition or rubber grommets between the threshold and underedge of the door.

3. Apply padding or caulking (putty) in the corners to seal them. Check for *no light* with the door closed.

4. Remove the threshold and recaulk under it. Embed the threshold in the fresh caulking and screw it tight.

Replace the Split Panel in the Door. To do this we must pry the rails away from the stiles. Most of the panel doors are pressure fit and seldom have nails holding the mortise and tenon joints together. Therefore, we can use wood blocks and a hammer to separate the parts and permit us to install a custom-made panel. If the door stile does not separate, then we must either find the nails and drive them out or assume that the door has glued joints.

If the joints are glued, we are left with only one solution. We need to cut away the molded edge on one side of the door to get the old panel out. This takes skill because no mistakes are permitted. After the new panel is installed, we must cut and fit new custom-made molding to hold the panel in place.

Replacing the Glass. Replacing the glass usually requires removing the molding on the inside of the door and the glass and old putty before we can make the repairs. Then we apply a bead of fresh putty by hand to the lip the glass will set in. We press the glass into the putty and install glass points to hold it firm. Then we nail the molding back in place and trim the excess putty from the outside with a chisel or putty knife.

Refitting the Door Jamb to the Door. Sometimes the only way to make the door fit well is to move the jamb. If the door hits the jamb, a simple trick is to place a wedge such as a 3-ft-long 2 × 4 between the door and jamb and force the jamb back against the framing. If this doesn't work, we must remove the casing and move the jamb. To move the jamb toward the door, we install wedges between jamb and framing. To make room for the door to close without striking the jamb, we remove the old wedges and make thinner ones. This permits the jamb to move closer to the framing and creates space or clearance between the door and jamb.

Apply the paint system. Applying the paint system depends on what we were required to do as carpenters. When we install a new unit, priming and top or finish coats have to be applied to the outside to match the colors there. The same can be said of the inside. But, even when repairs are made to the door and jamb, some priming and several top coats must be applied to make the job a quality one.

For the outside we should always use an exterior painting system. But for the inside we can use a semigloss latex enamel system. These paints cover well and withstand weather and wear and tear if quality paints are applied.

Concluding comments. Exterior doors can waste a great deal of energy when they fit and work poorly. They also create a great deal of frustration and anger when they do not close well or fail to lock properly.

The solutions frequently require the skilled carpenter; however, some owners have sufficient skill to make the needed adjustments or repairs. When in doubt as to what to do, always call on the carpenter or contractor. The money will be well spent and the job will last a very long time.

CHAPTER SUMMARY

The problems associated with restoring windows and exterior doors in a home may seem simple, but experience has proved that this is not the case. We have examined a wide variety of causes and effects that are problems. Many of these are well beyond the range of the owner to correct. In addition, we have learned that obtaining the exact replacement materials is not always simple. Many times the manufacturer has discontinued the line of product and made advanced products that are much better or simpler; thus parts are no longer available. The contractor is best suited to attempt first to locate replacement parts; if he or she cannot, the contractor knows how to have them made or can provide suitable substitutes.

Prices for the work can range widely. We learned that labor is intensive in most cases. Windows and doors must be removed, restored, repainted, weatherstripped, and the like. These activities take time and time is money. In addition, we must pay for the replacement materials. Not only are the materials expensive, but we also have to pay for their delivery and the associated costs of making estimates and pricing out the work, which includes a small percentage for office and equipment maintenance. Then, too, the contractor must make a profit or go out of business. Most contractors do not make much profit, some as little as 5% of sales. Regardless of how the expenses are arrived at, the windows and doors must be restored to working quality and weathertightness. This is important for both energy conservation and peace of mind.

6

WOOD DECKS, PORCHES, STAIRS, AND PATIOS

OBJECTIVES

To identify problems related to decay.
To understand the problems in restoring the deck, porch, stair, and patio framing.
To identify the techniques used in removing and replacing decking.
To identify the methods used to make repairs to stairs leading from the deck.
To use the proper materials when making repairs and restoration, including protective coatings.
To understand the overall construction technology for wood decks, porches, stairs, and patios with various underpinnings.

OPENING COMMENTS

Conditions or circumstances that require corrective actions. A wide variety of conditions or circumstances leads to the need for restoration of a wood deck, patio, or stairs. The most obvious is decayed or rotted planking used for the deck's surface. For the most part, treated materials such as 2 × 6 or the thinner 5/4 ×

6 pine are the most frequently used types. Figure 6-1 shows a typical deck. The chemical bath the wood is sent through forces the solution into the cells under high pressure. This penetration is fairly uniform, although in many instances the company quality control misses some that should be recycled. In addition, the chemical bath does not make allowances for or improve the wood where decay is already a part of the wood. Therefore, the contractor and carpenters need to be alert for treated materials that are not good quality and these should be set aside.

In residential deck, patio, and stair construction, creosoted timbers and other wood stock are seldom used. Materials protected by creosote are discolored because of the petroleum, smell bad, and cause burns in varying degrees when in contact with bare skin. They are most frequently used in and around wetlands, marshes, and along shore lines where water and tides make direct contact with the wood. Wood used above the high-water line is usually treated with chemicals other than creosote.

Rotted materials may also extend to the framing of the deck. Over time, water and waterborne insects can cause stringers, joists, and even posts to lose their capacity to support their loads. When this is the case, we have no alternative but to replace them. In this chapter, we will describe some of the techniques used.

Another problem with wood decks and stairs is that railings or bench railings work loose over time. Sometimes simply tightening the bolts cures the problem. But most of the time the wood around the bolts or screws erodes through weathering and pressure caused by occupants and guests. When this happens, we must either shore

Figure 6-1 Aboveground wood deck with perimeter benches.

up the materials with plates and larger bolts or screws or, better yet, replace the materials with new wood.

The stairs leading to and from the decks are also a major problem area. Any of the conditions mentioned above can extend to the stairs as well.

The porch connected to the house presents a slightly different set of problems. Porch decking usually is 1 × 4 fir or yellow pine. In the past and even today to some degree, the decking or "flooring" as it is commonly called was not treated. Even the framing for the porch was not treated in many earlier constructions. Where this is the situation, the owner must be made aware of the use of these materials and the cause and effect relationship with decay and infestation.

Contractor responsibility. As a contractor called to the home of the owner with a defective deck, porch, patio, or stairs, we would expect to find a serious state of disrepair. Minor problems will have been taken care of by most home owners, since this type of work appears simply as a remove and replace action and is certainly within the capacity of the able owner or friend.

The problem may be that the owner or friend did not exactly make the situation better. The replacement materials may not have blended in with the old materials nor were they fitted with quality craftsmanship. But, more than likely, the damage is structural and somewhat beyond the capacity of the owner or not economical for the owner to do.

We need to look over the structure and point out where the problem exists and make a thorough examination of it. Then we must relay the problems to the owner in terms of the repairs and restoration work needed to make the unit serviceable again. Any number of ways can be used to identify the problem. These range from showing the owner where the problem is to drawing a simple picture of the framing and pointing out what must be done, for example, to remove and replace a joist or two.

However, we need to do a lot more work before we can give the owner a fair estimate and contract price. So, we need to initially clarify the extent of work to be done by taking notes and having the owner agree in principle to these actions.

Homeowner's expectations. The owner expects a quality repair job at minimal cost. But he or she knows that both materials and quality workmanship are premium products, and the average costs are high for both. The owner further expects the job to be done quickly and efficiently. Finally, the owner expects the work to match the old design and architecture in style, color, and design, and he or she expects that quality materials will be used.

Scope or types of projects to solve. We will study four problem situations, progressing from the simple to the most difficult. In order they are (1) restoring the deck or flooring, which includes replacing the treads on the stairs, (2) restoring the railings or seat–railing assemblies, which includes the railings on the stairs, (3) restoring the framing and support members, which includes the stringer on the stairs, and (4) restoring the color and protecting the wood surfaces from weathering and infestation.

PROJECT 1. RESTORING THE DECKING OR FLOORING

Subcategories include replacing the 1 × 4 flooring on a porch; replacing the deck on an aboveground patio and deck and replacing the stair treads; fitting deck and flooring under columns and around posts; trimming overhangs; fitting pieces of tongue-and-groove between good ones; nailing flooring with face nails and toenailing.

Primary Discussion with the Owner

Problems facing the owner. The owner has called us in to perform the work of restoring the deck or porch floor because there were more problems to solve than just taking up the old boards and installing new ones. If we are involved with the deck of an aboveground wood deck or patio, there will also be a railing or perimeter seat arrangement to consider. If the job is a porch, there will be posts or columns to contend with as well as railings. These added features increase the difficulty of the restoration work.

Between the two, the deck is usually simpler to replace. The boards in the deck are not tongue and groove; in fact, they are usually spaced about 1/4 in. apart to allow water to readily pass through, whereas a porch floor is primarily made from tongue-and-groove fir or treated pine, and thus it is difficult to remove some of the pieces and replace them.

In the deck construction, the railings may either pass through the decking to the framing or be fastened to the deck itself. If the perimeter is seating, then some of the seat supports will pass through the decking and some will be fastened to the deck.

In porch construction, the posts supporting the railings may be fastened to the flooring, rather than pass through, but on some installations the posts do pass through the flooring. Where there are columns, these almost always sit on top of the flooring. And therein lies a problem, since the wood under the column frequently is rotted worse than elsewhere. These columns also hold up the porch roof.

Alternatives to the problem's solution. Few alternatives are available to the owner. The decking or flooring must be replaced where it is damaged. Either we replace all of it or just the pieces that absolutely require replacement. Our recommendation will depend on the extent of damaged decking or flooring and the probability that we can obtain an exact match of materials. Usually, decking materials can be found with the same dimensions. However, the T&G flooring may vary between yesterday's and today's standards or dimensions.

Statement of work and the planning effort. For this project, we can state the work to be done quite simply, such as *restore the deck flooring to a quality equal* *to its original condition and seal the new materials with a protective sealant.* If the project is a porch floor requiring replacement, we could say, *replace the defective, rotted, or damaged flooring and repaint to match the old color.* Nothing else needs to be added because of the simplicity of the job.

The planning effort for this project is also limited. We must determine the number of square feet of materials to buy and nails to use. We can also expect to secure all materials at the same builder's supply house, so one trip is all we will plan for. Unless the deck or porch is very large, we can complete the job in either 1 or 2 days. Then we will require another time frame for applying the sealant or paint.

Preparing the contract should not take long if the carpenters do all the work, including the painting. If a painting contractor is needed to paint the porch floor, then we will have to allow some time to receive his or her contract price. We might not have to consider the painting if the owner decides to paint the floor. But a note in the contract should stipulate this condition of the work.

Contract. The contract for this job should be a fixed-price type. There should be no unknowns at the time of its preparation. So, the body should state simply:

> For the price stated below, we will remove and replace the deck (or porch flooring) with replacement materials that are pretreated. The materials attached to the deck will be loosened, moved, and reanchored securely. After the decking is in place, we will apply one coat of sealant suitable for the exterior exposure. (The porch would be painted with a suitable porch paint.)

Material Assessment

Direct materials	Uses/purposes
5/4 × 6 Decking or 2 × 6 decking, treated	Replacement materials
Galvanized 10d or ring nails	Nail deck to joists
Clear, protective sealant or painting system for porch floor	Adds protection to the exposed deck
Match the color of the old floor	Match the color of the old floor
1/4-in. quarter-round	Replace old molding under overlap of porch floor (if there is molding)

Indirect materials	Uses/purposes
2 × 4 Blocks	Treated blocks where required for unexpected or expected added support
Screws or lag bolts	May have to replace rusted or broken ones while freeing railings, post, and the like
1 × 6 Pine or fir	May have to replace split column base materials (porch only)

8d Galvanized finish nails	Toenail railings and other materials freed up to replace the deck
Aluminum flashing	May need to replace at the point where the deck or porch floor contacts the house

Support materials	Uses/purposes
Carpentry tools	Construction
Workbench	Platform to cut on
Sawhorses and planks	Eases marking and cutting deck boards
Power saw and line	Simplifies cutting
Power drill and jig saw	Used for cutting around posts
Paint brush, roller	Apply paint
Step ladder	As required

Outside contractor support	Uses/purposes
Painter (optional)	Paint the porch floor and surrounding materials disturbed during the repair

Activities Planning Chart

Activities	Time line (days) 1	2	3	4	5	6	7
1. Contract preparation	X						
2. Scheduling and materials	X						
3. Deck or flooring removal		X					
4. Deck or flooring installation		X	X				
5. Sealant application			X	X			

Reconstruction

Contract preparation. The actual preparation of the contract will require the usual office activity. This includes obtaining the cost of materials, determining the time for workers on the job and travel time, and the standard application of overhead (fixed and variable costs). One trip to the owner for signing and setting up the time schedule will suffice to seal the bid/contract.

Materials and scheduling. As we see from the activities list, the entire job could take as long as 4 days for a moderately sized deck or porch. We will need fairly good weather; no snow or rain. On the first day the carpenters begin the removal of the old deck or porch materials. They might even begin installing the new materials before the day is finished.

The materials will have been delivered either the day work was scheduled to begin or the day before. This minimizes the possibility for theft and can, in some cases, save contract dollars because of discounts that apply.

Decking removal and replacement. Reconstruction begins with the removal of the decking or porch flooring. However, the carpenters must make a careful examination of the construction techniques used to install the railings, perimeter seating, or, in the case of stairs, the newel post, and, in the case of porches, the columns setting on the floor. Materials that are nailed to the floor must be loosened and freed. Let's discuss each situation separately.

Loosening the Supports of Benches. The legs of the perimeter benches are probably toenailed or screwed to the deck. See Figure 6-2. If screws were used, we are in luck, since we can use a power drill to extract them. Thus no damage will occur. On the other hand, we might find toenailing. This is more difficult to work with, since we must either drive them through or pry the leg up and cut them off. To drive the nails through, we need a very slim, tapered nail set. This shape avoids adding stress to the wood and minimizes splitting the wood of the leg. We can pry up the leg slightly, about 1/8 in., and then we can use a power jig or saber saw with a metal cutting blade to cut off the nails. Since there are legs about every 4 ft, we may spend several hours performing this job.

Normally, decking is cut around posts that support railings and banisters. Therefore, we should not have any trouble removing the pieces of deck around posts. Removing flooring under columns is another thing, though. For this job, we will use 2 × 4's and maybe small house jacks to take the weight of the roof off the columns. Then we must check under the bottom end of the column and locate the nails. Sometimes we need to remove added base molding first. Where possible, we cut the nails to free the column.

After the preliminaries are done, we can remove the decking. If all the deck must be replaced, we will use the same pattern that was used in the original design. If only a section of the deck is replaced, we will do essentially the same, but we will be more concerned about matching the spacing and joining pieces.

In the case of the porch floor, T&G flooring is used. Ripping up a section requires cutting through some of the damaged pieces to get crowbars and pry bars between boards. Installing the new boards is more difficult in patching than a complete new floor. To replace the final piece in a patch, either we have to trim off the tongue and wedge the last two pieces down together or we drive the last piece in from the end and trim off excess.

The ends of the deck must overhang the framing by 1 to 2 in. Since there are railings and or seating around the perimeter, or columns, posts, and the like, we may not be able to snap a chalkline and cut the excess at all, or we may be able to cut some of it. If this is the situation, we will need to establish the exact length of each piece before installing it. One way carpenters do this is to install a mason line from corner to corner. Then each piece is marked for length according to the position of the mason line.

End year construction

Detail of bench frames

Perimeter seating

Figure 6-2 Wood deck with perimeter benches.

After the decking has been replaced with new materials, the bench legs, columns, and other supports must be reinstalled with either nails or screws. Any other materials that were removed must also be reinstalled.

Applying the sealant or painting system. Most sealants used to aid in the preservation of exposed decks require a curing period of 30 days for the lumber to air dry and permit some of the chemical's properties to evaporate. Thus we need to return to the job in a month and apply a single coat.

If the porch floor is made of treated wood, we are obliged to wait the same time before applying deck primer and top coats of paint. However, if we used fir or pine that was untreated, we should not wait. Rather, we should apply the painting system as soon as possible.

Some characteristics we must look for in the paint are medium length soya alkyd resin coating, abrasion resistance, chip and flake resistance, high gloss, dirt resistance, and a stated exterior durability. The paint should also be heat resistant and have good hardness. The quality of the paint should moderately withstand alcohols, fuel oils and gasoline, animal and vegetable debris, lubricating oils, and various solvents without damage. Prolonged contact with these substances will affect the surface and eventually the base of the paint.

These paints require a minimum of two coats at 2 mils thick, for a total of 4 mils. Most can be applied with a spray gun, roller, or brush. Cleanup is with mineral spirits. Follow the directions on the paint can regarding temperatures at which to apply the paint and curing time between coats and between final coat and user use.

Several companies prepare clear coating to apply to treated wood. These liquids are very easy to apply and have a life expectancy of not more than 2 years. After that, the surface must be refreshed with another coat.

In cases where only part of the deck was replaced, a bleaching agent should be used on the old part of the deck prior to applying a seal coat. This product is available from numerous supply houses.

Concluding comments. The job of restoring the deck or porch flooring is not extremely difficult. Anyone with moderate carpentry skills and the proper tools can do the job. The difficulty is removing the other parts of the deck or porch that are fastened to the deck without causing damage.

Although this job could be the only problem with the deck or porch, there are frequently other problems as well. Where this is the case, the reader will combine examination of decking with one or more of the following examinations to make a complete job.

PROJECT 2. RESTORING THE RAILINGS OR PERIMETER SEATING

Subcategories include presecuring the railings, seat backs, seats, seat supports, newel posts, part of the railings, and stairs treads; replacing any of these items that are beyond saving.

Primary Discussion with the Owner

Problems facing the owner. Railings and sometimes perimeter seating around a deck or on the porch loosen from a variety of reasons. Figures 6-1 and 6-2 show two types of seating construction. Weathering is one cause, and it takes a long time before real evidence shows up. Weather causes several things that contribute to the problem. The first is the drawing action or pull from the sun on the nails or screws used in building the railing or deck seating. Over time the nails are literally pulled out of the wood by the sun's force. When this happens, the part that was held secure by the nail or nails is free to move.

The second cause from weather is the effects on the wood from changing weather. When the weather is wet, the wood cells absorb water and the fiber in the wood swells. When the wood dries out, it first gives up the water in the cells and later gives up the water in the fiber; this causes shrinkage. Since there are many different pieces in a railing or seating and they are of different sizes as well, the swelling and shrinkage occur at different rates. This pulling action loosens the joints, and the assembly feels and is loose.

The third cause is similar to the second but differs because of the way the wood is milled. Most wood is milled with a flat cut. Thus across the width of a board the grain is closer at the edges and wider at the center. Growth rings create the visible grain. Where the rings are closer together there is less warping, and where the rings are farther apart there is more. When some of the pieces in the seating and railing assemblies have wide boards (3½ to 5½ in.), warping takes place. The force of warping loosens nails or causes the nails to pull through some or all of the surrounding fibers. The result is a loosened railing or seating.

Weather also takes its toll on the nails and screws through oxidation. Even coated nails, bolts, nuts, washers, lag bolts, and screws, over time, loose their capacity to resist oxidation. When the condition progresses significantly, the fastener cannot hold the wood in place firmly.

Besides weather, outside forces may strike the railing and seating, such as children at play, people sitting and rocking on the railing, people jumping on the seats or over the rail, flying branches in violent storms, and the like. These forces cause the wood to move and break the solid bond created by the fastener.

Since the porch or deck was built as a unit, the conditions due to weathering generally prevail all over the unit unless there is some protection from the weather. Even though weathering takes its toll all over, the south and west sides always seem to be worse than the other two. This is understandable since the heat and rays of the sun are strongest from these two directions.

Alternatives to the problem's solution. Each part of the railing and perimeter seating can have several alternative solutions. However, they fall within several general categories, which are (1) replace the damaged, warped, or rotted pieces, (2) replace the entire assembly or unit, (3) tighten the loose pieces with new fasteners, (4) use a combination of categories 1 and 3, (5) maintain the style of the railing or seating, but add strengthening features to the design, or (6) select alternative materials more suitable to the climate and exposure while still preserving the architectural style.

The owner may know that there has been a long-standing problem with loose pieces in the assembly. He or she may even have noticed the changes in the utility of the seating or railings after long periods of rain and sun. So the contractor can offer alternatives 5 or 6, which will eliminate the current problem and also improve utility and quality. These two alternatives will add some cost to the restoration work, but the value added might be worth the costs to the owners.

Statement of work and the planning effort. For our generic project, we will state that the perimeter seating is loose in several places, the seat backs and seats are warped, and several are split and need replacement. Furthermore, the railing on one side of the staircase leading from the deck to the ground is very loose and partially damaged. Therefore, our first statement of work is to *remove and replace all split and warped seats and seat back materials and replace with treated materials.* We must also *tighten all bolts and screws that hold the seating to the deck.* This includes replacing the rusted ones and using larger ones where the wood, still serviceable, has weathered some. Then we are required to *restore the handrail on the stairs.* Finally, we need to *apply a chemical preservative to the new and aged wood.*

The planning effort began with the examination of the job site and the identification of the extent of the work of restoration. From that effort, we made a list of the statements of work. We also made a materials listing by measuring the sizes of wood used, types of fasteners used, and the dimensions of the assembly to gauge how much preservative to use. We also made mental notes about the worker-hours required for the work and the types of workers needed for the job. We also made mental notes about sources for replacement materials. Finally, we made an agreement with the owner to meet again and discuss the bid/contract offer and acceptance.

Contract. Since this work is fully exposed for examination, we can offer the owner a *fixed-price* type of contract. We can establish a firm listing of materials and these can be easily priced out. We can determine the worker-hours for carpenters and painter from empirical experience. Therefore, this is a straightforward contract. The body could be as simple as:

> For the fixed price shown below, we agree to restore the perimeter seating and railing on the left side of the stairs to their original condition. This work includes agreeing to remove and replace defective seating and railing materials, resecuring bolts and screws as needed, and recoating the entire decking, seats, stairs, and railings.

Material Assessment

Direct materials	Uses/purposes
2 × 6 Treated pine	Replace seats and seat backs
2 × 4 Treated pine	Replace seat caps materials
1 × 4 Treated pine	Railing materials

4 × 4 Treated pine	Replace newel post
Galvanized bolts	Replace rusted and corroded bolts
Galvanized lag screws	Replace the rusted and damaged screws
6d Common galvanized nails	Nail seats and other materials together
10d Common galvanized nails	Secure assembly members as required
X Gallons of preservative	Used to preserve the new and old deck wood parts
Mineral spirits	Cleaning agent for brushes and rollers
Galvanized drywall screws	Used in place of 6d nails (adds quality)

Indirect materials	Uses/purposes
Paint roller	Apply preservative
Stripping chemical	Bleach out the old materials to match the new materials (as required)

Support materials	Uses/purposes
Carpentry tools	Use for the reconstruction
Power saw and drill	Eases work
Sawhorses and planks	Work station
Ladder	As required to access the outside of the deck
Electrical extension cord	Power tools
Paint brushes	Apply preservatives in tight spots
Long drill bits	Drill through newel post

Outside contractor support	Uses/purposes
Painter	Apply the preservative

Activities Planning Chart

Activities	Time line (days)						
	1	2	3	4	5	6	7
1. Contract preparation	×						
2. Scheduling and materials	×						
3. Remove and replace the seating		×	×				
4. Remove and replace the railing				×			
5. Apply the preservative					×		

Reconstruction

Contractor support. The personnel at the contractor's office or the contractor herself will not require much support in the preparation of this contract since the parts of the job can be readily determined and stated in contract terms. The contractor would add overhead, transportation, and direct and indirect labor costs based on past practices. The materials estimate can be determined from a single call to the builder's supplier. An allowance for waste is usually added. Finally, the painter's bid is added to the contract price.

Materials and scheduling. The materials can be picked up the day of the job or they can be delivered on day 1 of the beginning of work. This day is set at signing of the contracts.

Restoration of the perimeter seating. The activity schedule shows that we allowed 2 days to restore the perimeter seating. Recall that two different designs were shown in Figures 6-1 and 6-2. We will follow this general sequence:

1. Begin the job by removing the defective materials.
2. Tighten all bolts and replace those that need replacement.
3. Install new seat back, seats, and cap pieces.

This is very straightforward work for the experienced carpenter. Many homeowners can do this work as well if they possess the skills needed. The idea in the removal stage is to avoid damage to serviceable parts. This means no hammer or crowbar marks, no chips or splits, and the like. Tightening the seat legs and back supports is the first logical step in the replacement effort. We need to make sure that every bolt and screw are firmly in place. The test is to try to move the wood assemblies. If they do not move, the foundation for the seating is good. Where supports cannot be tightened, they have to be replaced. We will use the old piece as a pattern to make the new piece. Finally, we install the new seat back and seat pieces and cap. Rather than nail the 2 × 6's (or 5/4 × 6 pieces) to the supports, we will use galvanized drywall screws. They hold better than nails and we do not have to bang in the nails.

Restoration of the railing. Figure 6-3 shows an example of the railing commonly used on stairs leading from the deck. The assembly consists of a newel post and railing assembly. The railing assembly is also called a balustrade. Included in the assembly are the top handrail, vertical pieces, and bottom rail.

The assembly is usually nailed together using the toenailing method. The newel post is usually bolted to the stringer to provide rigidity. There are many angles to contend with when new pieces must be cut and fitted. The experienced carpenter usually has no trouble dealing with this replacement, since tools are made for defining angles and marking them accurately. Hand cutting is usually used, and this requires considerable skill.

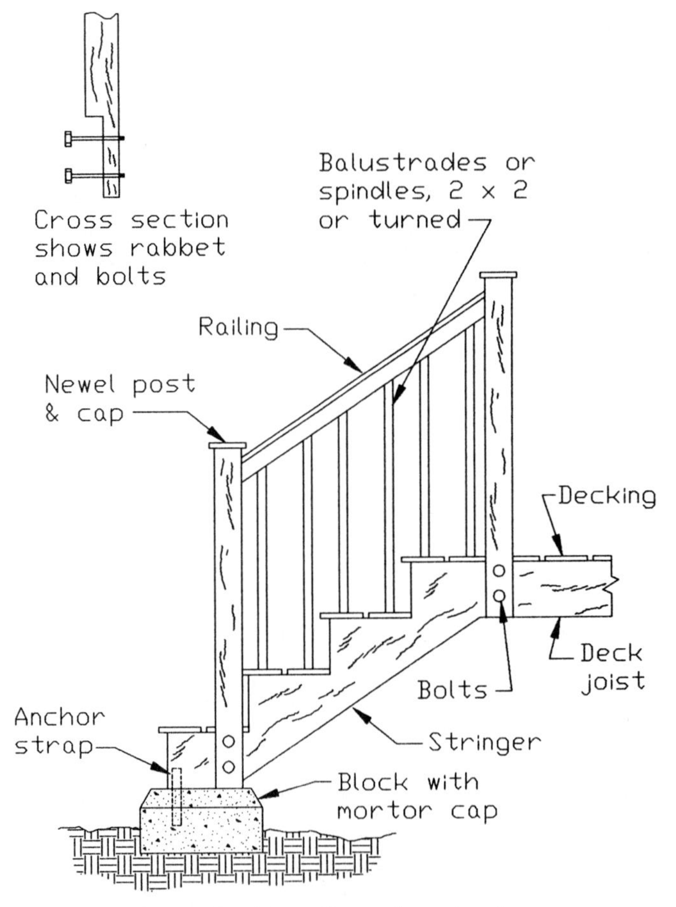

Figure 6-3 Stairs, railing, and newel posts.

The sequence of tasks we will use to replace the newel post and railing assembly is as follows:

1. Free the railing assembly from the two newel posts.

2. Remove the damaged newel post by either prying it loose or removing the bolts or screws.

3. Cut, fit, and install a new newel post.

4. Disassemble the railing system while trying to save some of the old pieces for models.

5. Using the old pieces for models, cut all the new pieces needed, and assemble the unit.

6. Install the new railing unit between the two newel posts by using the toenailing method. Set all nails below the surface.

Applying the preservative. The application of the preservative requires that the new wood, which is saturated with chemicals, be allowed to air dry about a month. This lets excess chemicals leach out from the cells and fibers.

Then the owner or contractor can apply a bleach to remove the gray or green discoloration from the old wood. A lot of the discoloration is caused by fungus and weathering. Once the deck is lightened and the time has elapsed, a single coat of preservative coats the surface and usually lasts up to 2 years. Then it must be recoated again. One gallon of preservative usually covers about 250 to 325 sq. ft of surface.

The preservative has a petroleum base; hence we must use mineral spirits to clean the brushes and roller.

Concluding comments. This part of an aboveground deck or porch would likely be the part a homeowner might try to restore using self-help. Overall the work is not that difficult. The pieces of lumber are not usually different from the standard cuts available from most builder's supply. With information as provided in this chapter, the tasks are executable.

But, when the owner lacks skill with carpentry tools, such as hand sawing accurately, using a sharp wood chisel, or accurately drilling holes in the lumber, then the better alternative is to hire a reputable contractor. This ensures that the quality will be restored to the decks perimeter seating and railings.

Coating the wood, old and new, with a clear liquid preservative is a simple task that one can do on a weekend or in an evening after work. The owner may opt for carpenters to do the repairs and then do the painting himself or herself.

PROJECT 3. REPLACING THE DEFECTIVE FRAMING

Subcategories include replacing defective joists, girders, headers, cross-bridging, sills, posts, and flashing; treating ground and building materials, footings and piers, stair stringers, and rafters on covered decks and porches; using joist hangers and other metal fasteners.

Primary Discussion with the Owner

Problems facing the owner. So far in this chapter, we have dealt with the parts of the deck and porch that are used by the occupants: the deck or porch surface and the railings and perimeter seating. This project focuses on the understructure that supports the flooring, seating, and railings. The discussions are also transferable to the deck or porch roof except for the differences in live and dead loads and the slope of the rafters and their required cuts. Recall that we discussed the rafter in Chapter 1.

The owner may face a wide variety of problems when the framing is involved. The list includes damaged or defective footings, piers, posts, headers, joists, stair stringers, and sills. See Figure 6-4.

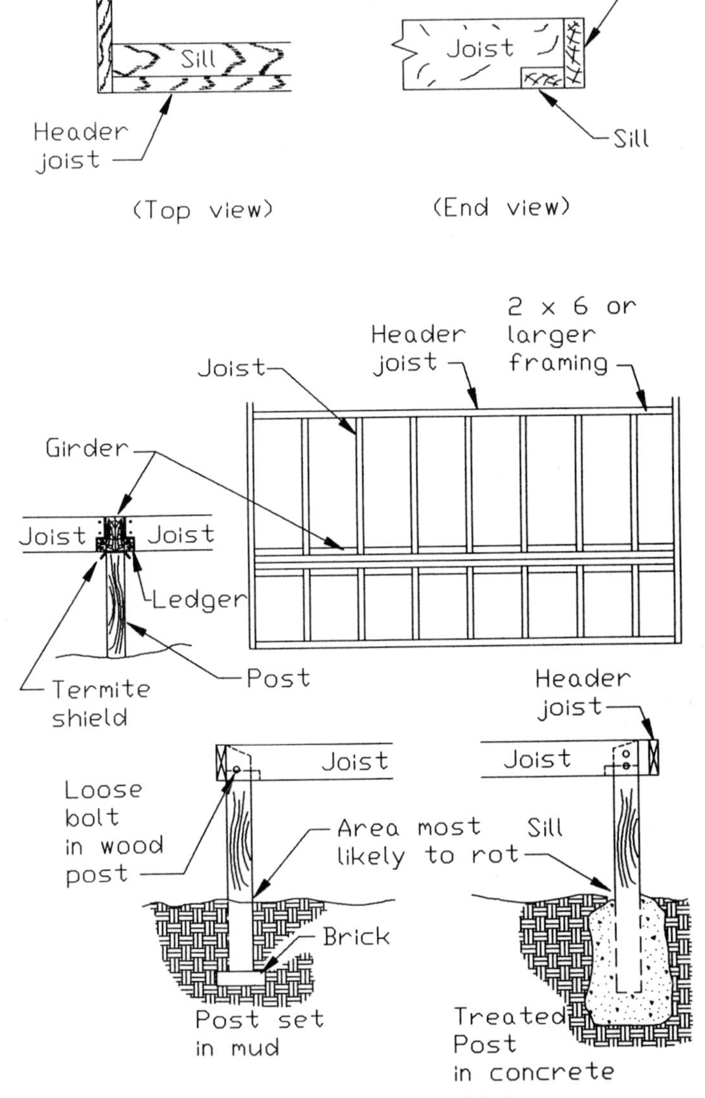

Figure 6-4 Framing of a wood deck.

Footings Can Shift and Settle. Most building codes require that footings be located below the frost line. Where the frost line is several feet down, it is very difficult for footings to shift. There must almost be a landslide or earthquake. But in many southern parts of the country, there is no frost line so the footings can be at ground level. In this situation the ground can shift when subjected to heavy rains, violent storms, and high winds, with the result that the footings shift.

Footings can also settle, even when they were set in virgin soil. The problem is more likely to occur in reclaimed ground and where backfilled ground is used.

Piers Shift, Crack, and Crumble or Are Broken by an Outside Force. Where the porch or deck is supported by piers, they are usually made from masonry products such as brick or block. Under normal exposure these units last indefinitely. But things can happen that cause them to lose their design characteristics. When this happens, a problem exists. A shift in the pier may crack the mortar joints or even crack the masonry unit. More often though the pier struck by an outside force ruptures. Fungus and constant moisture are detrimental to cinder blocks and cement blocks over time. Sometimes these materials must be replaced. The owner is entitled to an explanation that includes a description of the magnitude of the problem.

Treated Posts. Many decks today are made with pressure-treated lumber. These pieces range from the 4 × 4 to 4 × 6 to 6 × 6. In some locales, creosoted poles are used in place of treated lumber. Even these can rot and fail. They can last a very long time, but are also subject to decay, infestation, and rot. Posts are either set in concrete or mud. The area most likely to be damaged is the point where the post is flush with the ground.

Generally, posts have the problem of splitting. There are always internal tension and compression in every post. As the post is exposed to the weather, checks, splits, and even twists can occur. Usually the condition does not warrant removal and replacement. But when the condition is far advanced, we have an obligation to advise the owner that a major restoration problem exists.

Headers, Joists, Sills, and Stringers. Headers, joists, sills, and stringers are all made from nominal 2-in.-thick lumber. They may be as small as 2 × 6's to as large as 2 × 12's. These members in a frame are the horizontal pieces. Headers and sills form the box of the frame. The joists support the flooring, and the stringers support the stair treads and sometimes the risers when used. Even though these are treated members, they may rot, becoming infected with fungus where the preservative has weathered away or was very weak. In addition, these members are often cracked and broken by external forces such as we identified earlier. When these situations happen, we need to advise the owner of the effort involved in their replacement.

Girders. A girder, as shown in Figure 6-5, is customarily made from two or more 2-in.-thick pieces of lumber. A girder is used when the deck or porch is very large and some sort of added support is required to compensate for the span. A girder is also used when the deck or porch is made with a cantilever design. In cantilever design, a few feet of the deck "hang out in space." When stood on edge, all lumber can support a certain amount of weight at the end without bending. Architects sometime design the frame of the deck with a girder to minimize the need for support of the joists and headers of the deck's frame. A girder may have the same problems as any joists or headers.

Alternatives to the problem's solution. Each part described above can have one or more alternative solutions. However, there really are only several, which appear different because of the materials used. We can replace each damaged part.

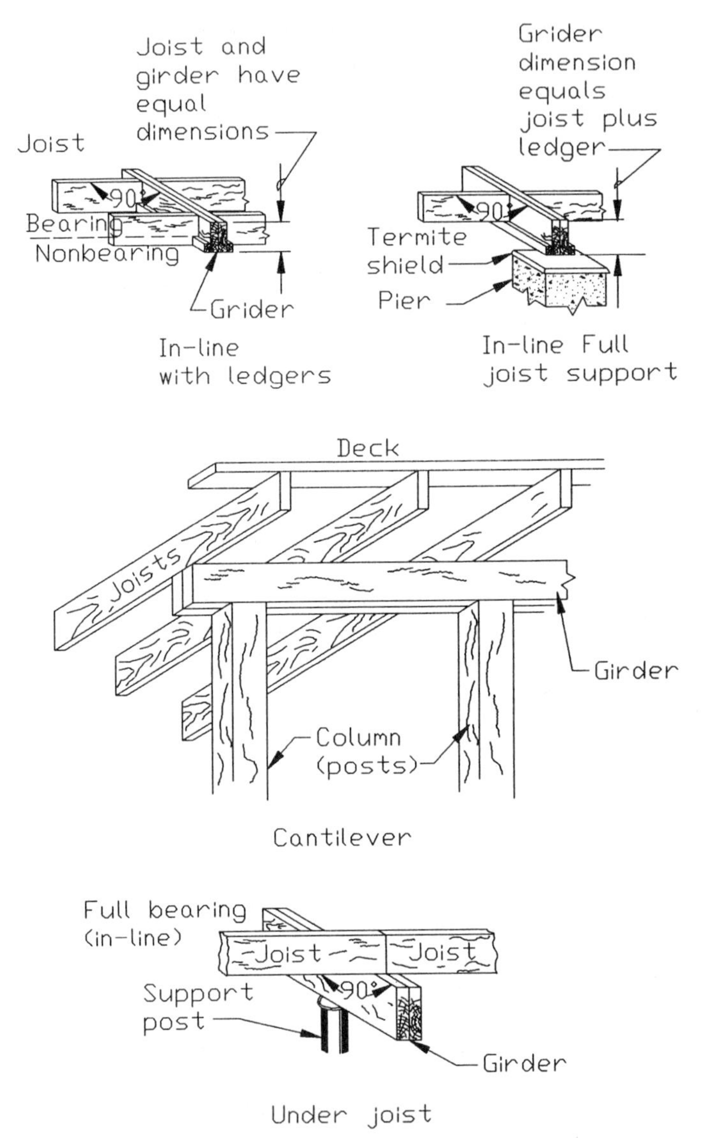

Figure 6-5 Wood girder used for large spans and cantilever deck construction.

If these are footings and columns or piers, we will pour new footings and raise a new pier from block or brick. We can remove any single piece of wood from the frame and replace it with a new one. We can remove and replace any post and replace it with a new one.

Sometimes we can splice new materials to the damaged ones after we have eliminated the cause of the damage. But this solution is very limited. We cannot use this solution where the spliced material is seen since it destroys the architecture of the deck.

Finally, we can sometimes reset the parts of the deck or porch that were moved. If parts are not damaged but have moved, then we can adjust them back to their original position and resecure them.

As we describe these problems with the owner, we need to show that getting to the damaged materials requires removing some of the decking and other parts, such as a railing. (However, in this section of the chapter we will only direct our attention to the framing segment of the deck.) We need to explain how the materials will be removed and the work needed to install the new members and post. These efforts usually require more than one worker because of their size and placement.

Statement of work and the planning effort. For this section of the chapter, we will assume that the problem was caused by physical forces that struck the deck and damaged a post and perimeter header and split several joists. The general statement of work would be to *restore the deck's framing and supports to their original quality condition.* More specifically, we will say, *remove and replace the damaged joists, headers, and posts with treated materials.*

The planning effort for this job requires the estimation of materials, which should be determined on the site at the time the contractor is called for an evaluation. We should determine the position of the deck relative to the house and check it for square, plumb, and level. Each post must be examined for alignment and soundness on the footing or in the ground. Finally, we should estimate the number of worker-hours required to complete the job. Since the work is outside, weather is a consideration in scheduling it. But, since the work is outside, the workers will not need to have the owner or a member of the family at home while the work is being performed. Even so, we will need a power source, and this may require access to the garage. We should expect that all the materials will be available locally and can be delivered to the site on the day work begins. The owner should expect a bid/contract within a day or two at the most.

Contract. This contract should be a fixed-price type. Every aspect of the job is visible. The owner has been briefed about the work that must be done, and our job is to put those items on paper and provide a cost. The body of the contract should state the following:

> The deck's framing and posts will be replaced where damaged, reset where moved but not damaged, and the completed work will replace the lost quality of the assembly. Special emphasis will be required to square the deck's frame.
>
> The materials and labor for this job are $××××.××.
>
> This price holds for 45 days from the date of this bid/contract.

Recall that fixed-price contracts do not allow for performing additional work nor adding improvements to the deck. They do not allow for added labor even if the workers take longer than expected.

Material Assessment

Direct materials	Uses/purposes
6 × 6 Post	Replace the damaged post
Bags of ready-mixed concrete	Pour around the post in the hole
2 × 8 Treated pine	Joists and headers
12d Common galvanized nails	Nail lumber together
16d Common galvanized nails	Nail lumber together
Bolts, galvanized	Replace bolts used in posts
Preservative	Coat new lumber after 30 days

Indirect materials	Uses/purposes
2 × 4's	Braces when squaring the deck
Shoring	Base for jacks

Support materials	Uses/purposes
Come-along	Tool used to force the frame to move
House jacks	Used to hold up the deck while the repairs are being done
Power saws and drills	Eases work
Carpentry tools	Construction
Paint brush	Apply preservative
Ladders	Used to access work above ground
Sawhorses and planks	Work station
Long extension cord	Power machines

Outside contractor support	Uses/purposes
None	

Activities Planning Chart

Activities	1	2	3	4	5	6	7
1. Contract preparation	×						
2. Scheduling and materials	×						
3. Remove and replace post		×					
4. Remove and replace framing		×	×	×			
5. Apply preservative					×		

Time line (days)

Reconstruction

Contractor support. Minimal contractor support is required of his or her office personnel since the contractor obtained the materials listing and determined the number of workers needed and their approximate time on the job. With these factors in hand, a standard contract can be easily prepared and priced. There are no subcontractors, so there is no delay in arriving at a total cost for the job.

Materials and scheduling. A simple materials list will be made in the office and called in to the builder's supply house a day or two before delivery is expected. This ensures that the materials will not lie in the open and be subject to theft.

Carpenters will be scheduled ahead of time and told that the job is expected to take not more than 3 days. After a drying-out period of 30 days, one of the carpenters will return to the job and apply a single coat of preservatives to the new wood.

Remove and replace the post. Removing and replacing a wood post that supports a deck frame is not a difficult task.

1. Install a jack under the deck and raise the deck slightly above the level point.

2. Remove the bolts securing the post to the frame.

3. Remove the post from the frame and pull it out of the ground. We usually need to dig around the post to loosen it and the concrete poured into the post hole at the time the deck was built.

4. Clean out the post hole and reshape it. If it is too large, we may need to build a form and set it in the hole.

5. Cut the post for length and set it in place. Plumb the post and use 2 × 4 braces to keep it plumb. Mix and pour concrete into the form.

After the deck materials are replaced and the assembly is squared, the post is fastened to the frame with bolts. If the deck was shoved out of square to the point where the deck interferes with the positioning of the post, we must perform activity 4, repair the deck frame, first.

Restore the deck framing. The largest part of the job is the work needed to restore the deck's frame and resquare the deck. We already established that the deck needed to be raised and supported by a house jack.

Resquaring the Deck. First, we can determine how bad the deck was shoved out of square. We use both the right triangle 3–4–5 technique and the × measurement technique. The × technique requires us to take tape measurements from opposite corners and compare the differences. In a perfect square the distances are exactly the same. When the deck is out of square, one measure is longer than the other, since we now have a parallelogram and not a square. When we use the 3–4–5 technique, we use the house as the base of the triangle, the side of the deck perpendicular to the house for the side, and the distance from the points as the hypotenuse. If the hypotenuse is too long or too short, the deck is out of square.

To move the deck, we may have to loosen some pieces, but more than likely the pieces are already loose. We then use come-alongs to move the deck back to square. While we have the come-alongs holding the deck square, we renail the frame and decking everywhere except where we must remove damaged members.

Replacing the Framing Members. We remove the members by freeing them from the decking, railings, and other members. The perimeter header may be parallel to the joists. Where this is the situation, it is quite simple to remove. But when the header is perpendicular to the joists, the joists are nailed to the header. This means that we must support all joists first. Then we can remove the nails (16d) with a nail puller, and, once all are out, we can slip the header away and cut a new one. Since the header was damaged, several joists were also damaged. We free the deck from the joists by prying up the deck from underneath. If that does not work, we need to either pull the nails from above or drive them through sufficiently to allow removal of the joist. Pulling nails that were set below the decking's surface will surely damage the surface of the decking. This would probably cause added expense. Driving the nails further into the deck with a nail set is a better alternative. Even if the nails are not driven entirely through, the joist can still be removed by twisting it and using crowbars from underneath.

We cut a new joist and install it. First, we nail it to the headers and then we nail the decking back down.

Finally, we lower the frame to level and fasten the post with bolts.

Coating the new wood. Coating the new, but air-dried treated lumber is done with a commercial coating and by brush.

Concluding comments. The work of removing posts and framing members is not a one-person job. It is also not a simple task that one should do unless the person has considerable experience with rough carpentry. We learned that special tools were needed to move the frame and support it while we completed other tasks.

If the framing is in a porch, then we will also have the problem of raising the roof and supporting it while the replacements are made. This will complicate the overall job.

PROJECT 4. PRESERVING THE NEW AND OLD MATERIALS

Subcategories include removing the stain from the deck; precoating treated materials; treating pressure-treated materials; painting pressure-treated and nontreated materials.

Primary Discussion with the Owner

Problems facing the owner. To extend the life of wood exposed to the elements, pine, for example, is pressure treated so that chemicals penetrate to the very core of a board or post. Poles may be pressure treated with chemicals or creosoted to extend their life expectancy. Sometimes, nontreated lumber and posts are used in constructing porches and decks, although not very frequently.

Even when these woods are treated, we know from experience that the chemicals leach from the wood over time. We also know that bacteria and insects will eat even treated lumber, and they readily chew away at untreated wood.

Some woods, such as cedar, red wood, and cypress, have natural agents that retard damaging bacteria and insects. But even these woods over time are infected and damaged.

The owner of a porch or deck is eventually faced with the unpleasant task of restoring the preservatives that kept the deck or porch from rotting or being eaten away. In the case of a porch, it is usually repainted every 3 to 5 years. The surface may look good, but it is the underside that is not easily treated. To help preserve the underside, we apply ground control poisons and ensure adequate ventilation.

On the aboveground deck, we also pay more attention to the exposed surface of the deck than the underside. The top or exposed surfaces generally are abused more and thus catch our attention. We only find out about the problems under the deck when we weed the flower beds or clean out the space once a year. Air circulation is also an important consideration for preserving the underside of the deck.

We can extend the life of the porch and deck by periodic applications of paint or chemical solutions. Some of the solutions contain stains; others are clear. All are available in discount retail outlets, paint stores, and builder's supply stores. Manufacturer's provide guidance for their application and the best conditions for good results. These generally include the following:

1. Weather and temperature conditions: 50° to 70°F, two or more days since last rain, no rain predicted for 24 hr.
2. Methods of application: brush, roller, and spray
3. Average square footage of coverage: 250 to 350 per gallon
4. Cleaning methods and materials: mineral spirits, good ventilation
5. Recoating time frame and time lapse between applying coats
6. Manufacturer's warranty

Sometimes the preparations include removing stains, fungus, mildew, lubricants, and fat. These conditions must be treated before applying the new coat of preservative. Fortunately, most are dealt with easily; some respond to household bleach or commercial products that clean and lighten. Other products, such as muriatic acid, toluene, and acetone, which are harsher and can damage skin and cause respiratory illness, must be used with proper protection and ventilation.

Alternatives to the problem's solution. The owner has a wide variety of choices in solving the problem of restoring the preservative to the deck or porch. If the porch or deck is painted, then a sound floor painting system is the best alternative. But a problem that must be solved is determining the type of paint currently used, roughing it up, and selecting one that bonds properly. A quality paint outlet can provide guidance. However, we should always, after sanding the surface, apply a bonding coat or prime coat to be sure that the top coats will not blister or peel.

Stains vary with their origin. For example, maple leaves leave a brown stain when wet and left on the wood. Oak leaves also stain, as do pine needles. Dropped food, left unattended, creates stains from the fats and natural chemicals they contain. Coffee, tea, and soft drinks create stains. Fungus and mildew create stains. Fortunately, all these can be bleached out with a mixture of detergent, bleach, and ammonia. We apply the mixture with a corn broom and rub it into the wood's surface. After letting it set about 30 min, we wash off the deck and apply more to the remaining stains.

Petroleum, paint, and other more difficult stains and discolorations require the use of stronger chemicals. These chemicals must break up the stains at or near the molecule level. Many of these chemicals can ruin paint, discolor adjacent surfaces, bleach vegetation or kill it, and cause severe burns if the spray hits the surface. But they do the job.

Statement of work and the planning effort. For this section of the chapter, we assume that the deck has been in use for several years. There have been many parties, kids played on it with toys, some construction and painting chores were done on it, and thus it needs attention. Our statement of work begins with *restore the deck's surface and exposed areas, including railings and steps, to a lighter color and preserve the surfaces.* Some added detail should also be included, such as *remove the stains, paint, and ensure that the scratches and gouges are appropriately treated.* Also, *remove stain caused by nails and screws and apply a coat of polyethylene varnish to these areas to retard further rusting.*

The planning effort is really simple and easy to carry out. The estimate must be prepared and an offer made. Then the materials are picked up from the supplier. Finally, workers are assigned, and they must complete the work. The only problem that should have bearing on the job is the weather.

Contract. This contract should be prepared by a painting contractor who has the knowledge and experience to perform all statements of work. However, we could also employ a general contractor who would subcontract the work to a painter, or if the contractor had a full-time painter on the crew, he could have the work done by his crew.

This contract can be a fixed-price type. There are no hidden areas that could cause added costs. All problems that need solving are fully visible and can be easily identified. The contractor or painting contractor should measure the dimensions of the deck and estimate the square footage of railings and stairs. Data collected at the first inspection must also include the type of stains, percentage of surface covered with stains, and the type of solutions needed to clean the deck and prepare it for applying the preservative.

The contractor needs to talk to the owner about the number of coats of preservative to apply, as well.

The contract should state:

> We agree to strip the exposed wood surfaces of stains, natural discolorations, and overspray paint. Then we will apply two coats of clear preservative that have an expectancy of 2 to 3 years or as long as specified by the manufacturer of the product. Furthermore, we will coat the nail, bolt, and screw heads in the railings and stairs, headers, and the like, where there is evidence of rust to retard the formation of more rust and stains.

Material Assessment

Direct materials	Uses/purposes
Chemical preservative	Top coat of wood surfaces
Mineral spirits	Cleaning solvent
Muriatic acid	Clean stubborn stains
Bleach or bleaching solvent	Eliminate mildew and fungus
Detergent and ammonia	Clean away food and drinks (sugar)
Rust remover	Clean away stubborn rust streaks
Sandpaper	Smooth rough spots

Indirect materials	Uses/purposes
Roller heads	Replacement heads
Paint brush	Apply solvents, chemical baths, and top coat
Roller tray	Stores the fluid during painting
Rags	Wipe spray and spills
Water	Mix with detergent and muriatic acid
Brooms and mops	Apply cleaning chemicals
Buckets	Mix detergents and also muriatic and water
Gloves, mask, long-sleeve shirt	Safety clothes while dealing with acid

Support materials	Uses/purposes
Drop cloths	Cover shrubs and walks
Plastic sheet	Cover adjacent walls to avoid splashing

Outside contractor support	Uses/purposes
Painter or none	Perform the work (none if contractor is the painter)

Activities Planning Chart

Activities	Time line (days) 1	2	3	4	5	6	7
1. Contract preparation	X						
2. Scheduling and materials	X						
3. Cleaning and preparation		X	X				
4. Top coating				X	X		

Reconstruction

Contractor support. The preparation of the contract at the office should take less than 1 day. The office manager, estimator, or contractor's secretary will probably prepare the documents and obtain the prices for the materials. This person will also have the percentages for fixed and variable costs to add to the price. The contractor may not even have to calculate the quantities of materials from the materials listing, since the office person can do this from the data collected at the site and knowledge of amounts needed from past experiences.

Materials and scheduling. For this job, most of the materials can be obtained from a paint supply house, and the rest can be obtained from a variety discount store. These materials do not require a great deal of room; therefore, they can fit into the worker's truck or car. Thus they can be picked up on the day scheduled for the job to start.

Notice that 4 days are allocated for the job. The first 2 days are used for cleaning and bleaching the wood. The last 2 days are for applying the top coats of preservative and sealing the nail, bolt, and screw heads to prevent further rusting.

Cleaning and preparation. As discussed in earlier paragraphs, there are various steps in the cleaning process:

1. Protect the shrubs and wall from spills and splashing liquid.
2. Clean with detergent, bleach, and ammonia.
3. Clean lubricants with solvents.
4. Clean paint spray with muriatic acid and water.
5. Clean stubborn rust stains with rust remover.
6. Make a final wash with clear water.
7. Smooth rough surfaces with sandpaper.
8. Let dry.

Apply top coats. After the cleaning is completed and the wood has dried, the first top coat of preservative should be applied. This can be done with a brush around railings and steps and with a roller on flat surfaces. Some products also permit the use of spray guns. Since the liquid is so thin, the workers must be fairly accurate and careful to ensure that all surfaces are adequately covered. Some instructions require the wood to be saturated, and then excess fluid is brushed to other spots. We expect to use 1 gal of preservative for every 125 to 200 sq. ft.

After the top coat has dried the appropriate time, usually overnight, the second coat should be applied. This coat will cover more square feet per gallon than the first. We expect to use 1 gal of preservative for every 275 to 350 sq. ft.

After applying the second coat, we need to use a very small paint brush (1/4 to 1/2 in.) to apply the polyethylene over the nail, bolt, and screw heads in the railings, stairs, and vertical members of the framing.

Finally, we need to remove the drop cloths and plastic sheets.

Concluding comments. This project can be done by many homeowners. Therefore, the owner needs to make a judgment about the cost of a do-it-yourself job or contracting it out. It will take no fewer than 4 days to do the job, and maybe longer when we add in the time to buy all the materials in the materials assessment. It might be cheaper to contract for the work if the owner has a well-paying job.

Only a few parts of the job involve personal danger. Those are when we use the acids and solvents. The rest of the job requires little lifting and strain.

CHAPTER SUMMARY

We have divided the reconstruction of a deck or porch into four basic parts. We have identified and shown how the problems with each part can be solved and the amount of effort and materials needed to complete the restoration.

However, in many real-life situations, there are times when more than one section of this chapter is required to completely restore the deck or porch. When that situation exists, the owner or contractor can use the guidance from each section to produce a single solution and contract.

7

CONCRETE SLABS, SIDEWALKS, AND DRIVEWAYS

OBJECTIVES

To understand the methods of constructing concrete slabs, sidewalks, and driveways.
To identify some of the causes for decay and deterioration of the concrete slabs, sidewalks, and driveways.
To learn how to deal with small problems and large problems.
To examine some of the types of contracts suitable for restoration of the concrete work.
To examine material assessments to appreciate the variety of materials that may be required for a restoration project.
To obtain an understanding of the use of labor and the types of skills needed for this type of work.

OPENING COMMENTS

Conditions or circumstances that require corrective actions. Concrete slabs, sidewalks, and driveways have a lot of similarities, but there may be some differences. The similarities begin with the construction of a form to contain the concrete and give it shape and form. Sidewalks and driveways have expansion joints strategically placed. See Figure 7-1. These joints are used to allow the concrete to expand and contract with predictable results. Soon after the concrete dries and cures, the concrete below the joint cracks. This is where we want it to crack, since the crack is not visible to the eye. If the joint were not made or there were not enough of them, cracks would occur randomly across the concrete. Figure 7-2 shows a driveway that is 20 years old and has severe cracks due to expansion.

Concrete slabs usually do not have expansion joints and eventually may crack just like the driveway. Sometimes the slab is built with a perimeter footing that is reinforced with rebars. This perimeter of steel strengthens the slab considerably. Still, slabs can crack, but often from other causes than internal stress.

Concrete that cracks at expansion joints does not present problems other than that weeds grow out of the crack. Expansion cracks other than at expansion joints cause problems over time. They are unsightly, weeds grow in them, one side of the crack sometimes rises higher than the other, and, eventually, the crack enlarges to a gap zigzagging across the concrete surface.

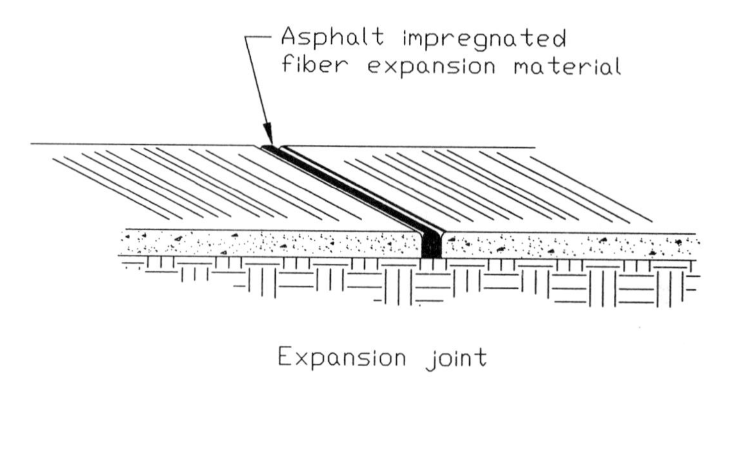

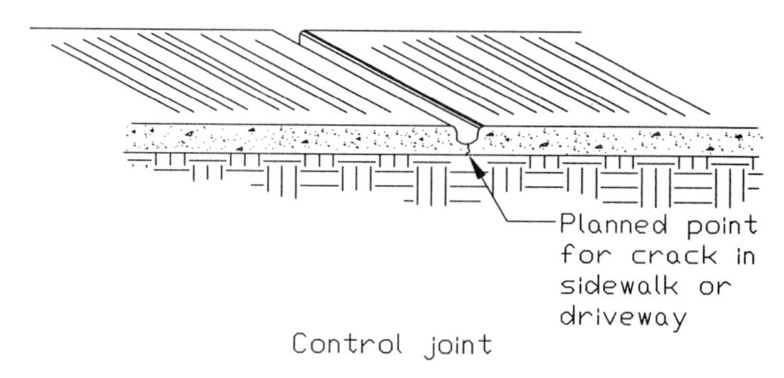

Figure 7-1 Expansion joint in sidewalk or driveway.

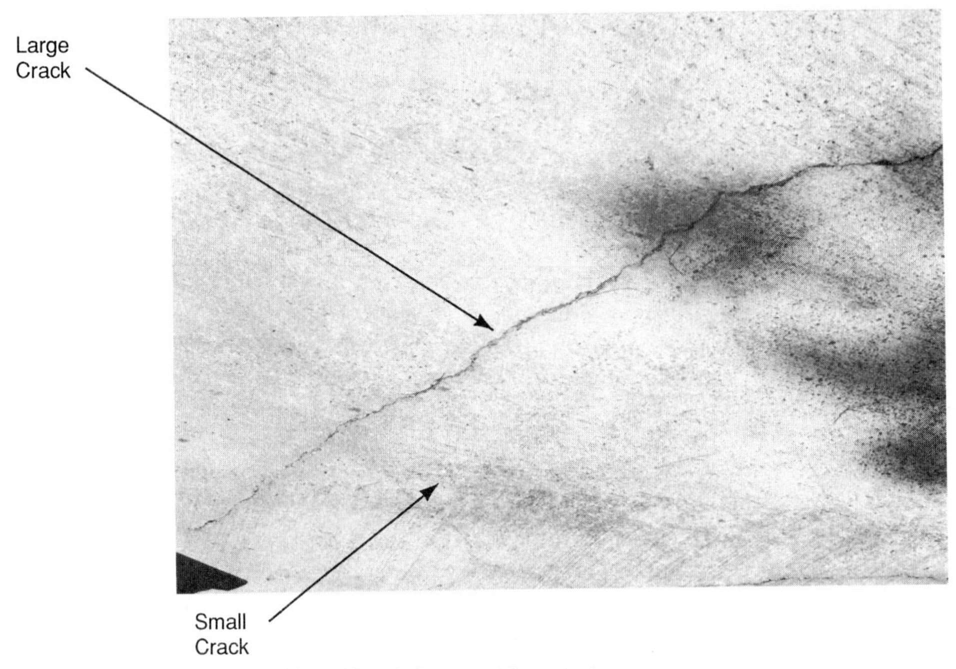

Figure 7-2 Driveway with cracked concrete.

Concrete is initially finished in a variety of ways. The common ones are (1) very smooth, which is done with a metal smoothing trowel, (2) the broom finish, which is slightly rough and done with a stiff-bristled broom, and (3) a very slight roughness created by using a darby or float on the surface.

An exotic way of finishing the surface is with the use of rock salt. This produces a pitted finish. The surface can also be covered with small stone from which a lot of the cement is washed away to leave the exposed aggregate. The last two methods are more costly than the first three.

These various finishes abrade away over time from a variety of causes that leaves the finish looking deteriorated and unsightly. The wearing generally is caused by the action of weather. Heat causes swelling, ice causes cracking, rain causes erosion of cement and small aggregate or sand. The abrasion is never uniform, and it may not be apparent for many years. But one day it's there, and we have a problem.

Scaling is another cause for concern. Scaling is the breaking away of the hardened concrete surface to a depth of about 1/16 to 3/16 in. This happens most often shortly after the concrete has been poured. However, scaling can result from repeated freezing and thawing and from using deicing salts.

Crazing, which is similar in appearance to the cracks in an egg shell, results from surface shrinkage shortly after the concrete has hardened. The cause is usually too rapid a drying action.

Contractor responsibility. When the general contractor or mason contractor is called to the job site by the owner, an examination of the concrete must be made. The contractor must first determine the different problems and then discuss them with the owner. He or she should explain that natural cracks occur at the control joints and that if the problem of weeds growing between the cracks is a serious problem, liquid bituminous can be poured into them after they are cleaned thoroughly. But the joints must remain an essential part of the sidewalk or driveway.

Cracks outside the control joints are a serious problem, as the contractor will state. They go all the way through the concrete. They usually are caused by pressure from above or underground. Heavy trucks, such as a concrete delivery truck, can crack 6-in.-thick driveways in many cases. Under some conditions, even cars can crack the driveway. But more often the crack happens because something under the concrete changed. Tree roots grow under the slab, sidewalk, and driveway. These cause a lot of damage. They can lift whole sections, crack the concrete, and permeate the ground under the concrete so that it gives just enough to permit cracking. Underground streams and high water tables also permeate the soil and create the softness needed for slight give of the concrete and a crack results. When cracks are large and uneven, we can sometimes raise the concrete and remove soil and lower it to make it level. When there are many in close proximity, the only cure is to replace the slab, driveway, or sidewalk.

Crazing or scaling are different problems and require different solutions. These two conditions are seen soon after the concrete has been poured. This means that the concrete unit may still be under warranty. The contractor may tell the owner that the problem happened because of improper drying and curing conditions, improper handling of the mixture, or a poorly mixed batch of concrete. Any of these answers could be correct, but the problem for the contractor and owner is how to correct the situation. The contractor can etch the surface with an acid bath and remove the worst scaling and then apply a top dressing of concrete and float it. Or the concrete may need to be broken up into pieces and removed. Then new concrete is poured and finished under stricter controlled conditions.

Homeowner's expectations. The owner knows that the problem is getting out of hand and something must be done. He or she would like a very simple and inexpensive cure to the problem. After the job is done, it should blend in with the older concrete or adjacent surfaces that have aged. The owner expects any newly poured concrete to match the color of the older concrete that was serviceable and did not need replacement. These are serious expectations for the contractor. They are even more serious if the owner tries a do-it-yourself method. All restoration work will be expensive.

Scope or types of projects to solve. In this chapter we will first solve the problem of a raised concrete sidewalk. Then we will solve the problem of replacing the slab and adding reinforcement. We will also replace the driveway with serious cracks by adding properly located control and expansion joints. During this discussion, we will also include a note on blacktopping the driveway after patching the serious cracks.

PROJECT 1. RESTORING THE SIDEWALK

Subcategories include forming the sidewalk; preparing the subsoil; selecting the proper climate conditions for pouring and finishing; selecting and creating the finished appearance; estimating the amount of concrete required.

Primary Discussion with the Owner

Problem facing the owner. The owner has a sidewalk about 50 ft long. It starts at the rear of the house and extends around the south side about 3 ft from the wall. On the stretch on the south side, a tree that has grown quite large over the past 15 years. Its roots have grown and cracked and lifted the sidewalk in two places about 12 ft apart. The concrete has weathered to a gray and is stained naturally. There are also smaller cracks at several places near control joints, and some few areas have erosion revealing the stones (aggregate). Although there are many problems, the only one that poses a safety hazard is the concrete lifted from the pressure of the roots.

Should the owner have only the broken and damaged sections of the sidewalk replaced, or should the entire 50 lineal feet be replaced?

Alternatives to the problem's solution. The alternatives are few. The entire sidewalk can be replaced. This ensures that the color, texture, and finish will be uniform. Replacing the dangerous sections will eliminate the safety hazard, but the color will likely not match. The texture and finish will be fresh compared with the old and weathered sidewalk. The broken pieces near control joints can be replaced with fresh concrete, but the color will not match. When cost is the primary factor and means more than architectural restoration, the owner will need to make this known early in the discussion about repairs or restoration.

Statement of work and the planning effort. We recommend that the entire sidewalk be replaced to enhance the value of the property and minimize further costs as more of the sidewalk deteriorates. So the statement of work should be *remove and replace the entire length of sidewalk, and while doing the work, remove the reason for the destruction of the sidewalk caused by underground and ground-level roots.* The statement of work should be more specific to include *the finish will be either floated or broomed.* There can even be a specification pertaining to the color, type, and depth of the concrete and the width of the sidewalk. These could be *the color of concrete must be white, with a mixture of 2500 lb per cubic yard, and equal to the thickness of a 2 × 4. The uniform width of the sidewalk will be 4 ft wide.* Finally, we need to have a statement describing the cleanup and condition of the grounds on both sides of the sidewalk after the forms have been removed. This reads *all residue of concrete and runoff as well as forms will be removed from the ground and not tilled in. Furthermore, ground on both sides of the sidewalk will be leveled and filled with top soil or sod as required. Gardens between the sidewalk and house will be restored to their original condition, which may require new ground cover and fertilizer.*

The planning effort begins at the job site, where we make an assessment of the work to be done. The discussion with the owner needs to determine who will do the landscaping repairs. If the owner does it, the contract must clearly state this. If, however, the contractor will do the landscape restoration after the new sidewalk is installed, then the planning and note taking must assess how much this work will take in terms of labor and new materials. We also need to plan who will do the gardening. Will the workers be the masons or a subcontractor?

We also need to determine how long the entire job will take to remove the old sidewalk, set forms, and pour and finish the 50 lineal feet. There will be transportation costs with trucks and labor to cart off the old concrete; these must be assessed as well.

We also must remove the problem of roots, which were part of the cause for the sidewalk's destruction. Normally, this work would be done at the time we set the forms and the masons would remove the roots. The work will probably add 1 or 2 hours to the labor costs.

Contract. Whether it is one contractor or a contractor and subcontractor joint effort to bid on the job and later perform the work, we should be able to prepare a fixed-price type contract. The well-developed contract will contain the statements of work. The body of the document could be similar to the following:

> For the fixed-price stated below, we agree to remove all the old concrete sidewalk from the premises, eliminate the immediate cause of the damage to the old sidewalk, install a new sidewalk 4 ft wide, 3½ in. thick and use a 2500-lb pour that includes white cement. We further agree to repair the landscape on both sides of the new sidewalk with sod and/or ground cover if required.
>
> Total cost $×××.××.

Material Assessment

Direct materials	Uses/purposes
Concrete	Replaces the old damaged concrete
Top soil	As required for restoring grounds
Ground cover	As required for restoring grounds
Sod	As required for restoring grounds

Indirect materials	Uses/purposes
Forms	2 × 4's or Metal forms and stakes and nails

Support materials	Uses/purposes
Mason and carpentry tools	Construction
Wheel barrow	Move concrete
16-lb. Sledgehammer	Break up old concrete
Truck	To haul off waste materials
Rakes and hoes	Move concrete and perform grounds repair

Darby	Level concrete mixture
Pneumatic jack	Break up concrete
Power machine	To lift up pieces of sidewalk and place them in a truck

Outside contractor support	Uses/purposes
Landscaper	As required if a subcontractor

Activities Planning Chart

Activities	Time line (days) 1	2	3	4	5	6	7
1. Contract preparation	×						
2. Scheduling and materials	×						
3. Prepare the forms		×	×				
4. Pouring and finishing sidewalk				×			
5. Restoring grounds					×	×	

Reconstruction

Contractor support. The office personnel will prepare the contract and in doing so will obtain the bid/contract price from the landscaper. They will also perform the estimates for concrete, forms, and labor. Then they will apply the overhead costs and allowance for profit. One call they will make is to the local producer of ready-mix concrete for a price on 2500 lb of mix using white cement. If the mason contractor has metal forms, these can be used in lieu of 2 × 4's and wood stakes, and a rental/use fee will be included in the worksheet for the job.

The entire time for this preparation should take less than a day after they receive the bid from the subcontractor.

Materials and scheduling. We know that there will be a need for concrete. The sidewalk is planned to be 4 ft wide by 50 ft long with one 90° bend at the corner of the house. Figure 7-3 shows this. We need to determine the amount of concrete needed for the job.

Fortunately, we have the necessary information to make the decision. Concrete is prepared and sold by the cubic yard in ready-mix form. This translates to 36 in. by 36 in. by 36 in. = 46,656 cu. in., or, stated in cubic feet, 27 cu. ft. All we need to do is determine how many cubic feet or inches we need for the job.

Problem solution 4 ft wide × 50 ft long × $\frac{1}{3}$ ft thick = 66.67 cu. ft. We would round this up to 67 cu. feet.

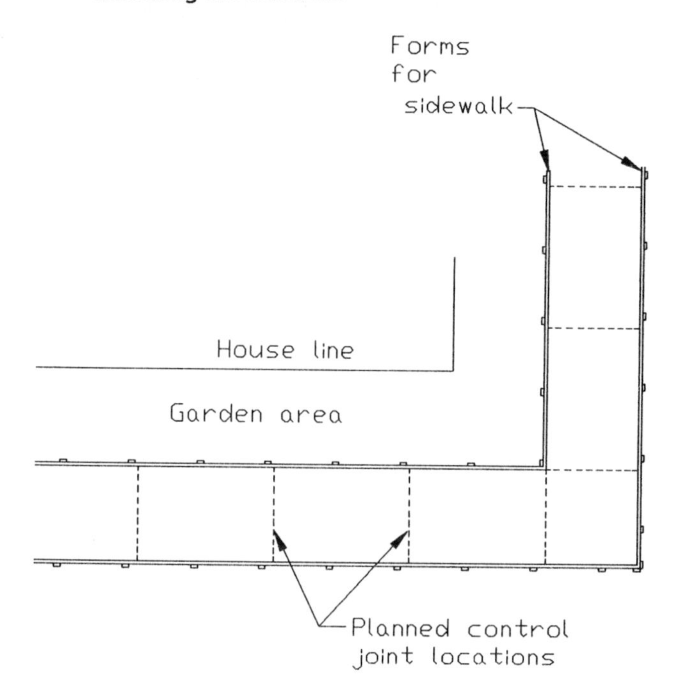

Figure 7-3 Sidewalk layout with 90° bend.

Then we divide the requirement by cubic feet per yard to obtain the number of yards needed. 67 cu. ft/27 cu. ft = 2.48 cu. yd. We then round this up to place an order for 3 cu. yd.

The materials for the forms are equally easy to calculate: 50 ft + 50 ft + 8 ft for ends = 108 ft. The number of wood stakes, either 1 × 4's or 2 × 4's is one per 3 ft of form plus extras of the ends. This totals to 108/3 = 36 + 4 for ends = 40 at 2 ft long. So we would buy 80 ft of materials.

The schedule must include the concrete company, landscaper and contractor's workers as well as the agreement by the owner. This is easily accomplished by the office worker or manager or by the contractor. We see from the activity list that the entire job will take a maximum of 5 days from the signing of the contract or start date agreed to by all parties.

Prepare the forms. Preparing the forms is the last in the series of tasks required for this phase of the project. One of the hardest jobs will be taking up the old sidewalk using a sledgehammer and manually throwing the pieces into a truck. Several power tools are available and include pneumatic jacks that break the concrete into small pieces and small machines such as the Bobcat, which can dig and lift sections of the sidewalk. No matter how we do it, the first task is to remove all the old concrete sidewalk.

Then we would do the following:

1. Chop out all roots along the path of the sidewalk to a depth of 6 to 8 in.

2. Rake the soil to the approximate contour of the ground. This can be anything from flat to a tiered approach.

3. Place the form pieces closest to the house first. This makes it easy to maintain perfect alignment with the house.

4. Place the opposite forms 4 ft away. We usually use a spreader stick or two to maintain the separation.

5. About every 12 ft, nail a spreader piece of 2 × 4 or 1 × 4 to the form to aid in stabilizing the form when the concrete is being poured and moved around.

6. Verify that the form meets the following standards:

- **a.** It is staked about every 3 ft.
- **b.** It is staked at the corners and at joints in the form.
- **c.** Its position is accurate.
- **d.** Opposite forms are level.
- **e.** Its spreaders are in place and nailed with double-headed nails for easy removal.

7. Smooth and backfill dirt inside the form to make the depth even with the bottom of the forms.

8. Spray the dirt with water to settle it and tamp it firm.

9. Clear the areas on both sides of the form to make a safe work surface and to ease the screeding, darbying, and finishing tasks.

This is all that is required for this phase, yet there is a great deal of physical activity and stress associated with the work.

Pouring and finishing sidewalk. The concrete truck will arrive on the site as planned. The driver does not expect to remain there over 30 min. This means that we have to move 2½ yd of concrete from the truck into the form quickly. The truck is equipped with chutes that aid us in completing this task. However, if the truck is blocked from reaching all parts of the sidewalk forms, we must wheelbarrow the concrete. Sometimes the driver will help out with the pouring, but we will need to make sure that enough workers are on the job if a lot of wheelbarrowing must be done.

Once the concrete is distributed throughout the form, we proceed as follows:

1. Screed the concrete flush with the top of the forms.

2. Eliminate pockets of air alongside forms with a 1 × 4 using an up and down arm movement. This action causes the cement and sand in the concrete to move to the form and forces the large aggregate (gravel) back into the mixture. Even though the sidewalk's edge will be covered by sod and ground cover, its strength is increased by doing a good job of this task.

3. Remove the spreader pieces after the concrete is screeded and tamped along the edges. Fill in the gap with excess concrete.

4. Let the mixture set until the top dulls and feels firm.

5. Darby the concrete.

6. Float the concrete.

7. Use the edger tool to round the edges against the forms.

8. Make 4-ft spaced pencil marks and use the control joint tool and straightedge to complete all joints.

9. Protect the concrete. If the weather is well within the range for proper curing, the protection can be simply blocking off access. If the weather is near the freezing point, we must provide a heat-generating cover. This could be hay sprinkled over the surface or a sheet plastic draped over the sidewalk with a heater installed under the sheeting. Plastic sheeting is also used if there is a chance of rain or snow. In very hot weather, we must add a fine spray to the concrete to keep the thermal action active. If it stops too soon, the entire batch will have a very short life span.

The concrete cures at the following rate under near ideal conditions. It can be walked on after 1 to 2 hours to complete the finishing work. It dries hard in 24 hours. It takes 1 week to cure to 90% of its final strength, and it is fully cured in 30 days or slightly longer.

Restore the grounds. After 24 hours, we remove the forms with care to avoid making any mars or cracks in the uncured concrete. The forms must be cleaned with water and a stiff bristle brush of residue cement and concrete. All nails must be removed and placed in a waste container. The forms should be loaded on a truck at this time.

Using shovels and rakes, we must remove all spills from both sides of the sidewalk in accordance with the contract specifications.

Then we begin the relandscaping activities. These include (1) adding top soil where needed, (2) adding sod alongside the sidewalk, and (3) restoring the gardens from house to sidewalk.

Concluding comments. Many owners can perform all the tasks of replacing a sidewalk. This project has illustrated the amount of work involved. We showed how strenuous the work is as well as the need for accurate quality workmanship. When the project gets beyond 1 yd of concrete, two people are really needed for the pouring and finishing work. Tearing up and carting away the old sidewalk is very physical. It takes a strong person who usually lifts weights or weighted objects at work to swing a 16-lb sledgehammer and break up the concrete.

We mention these cautions to make sure that you are aware that the work is hard, and must be done quickly, and, finally, lots can go wrong.

PROJECT 2. REPLACING THE SLAB WHERE REINFORCEMENT IS ADDED

Subcategories include using rebars and wire mesh; using a footing/slab combination pour; forming the perimeter; floating and finishing with a steel trowel.

Primary Discussion with the Owner

Problem facing the owner. When the house was first built, the contractor poured a patio slab at the rear of the house using 2 × 4 forms and no reinforcement. Over the years the concrete has weathered, but the real problem is with the crack in the floor and the broken corner. Obviously, the contractor saved materials and labor while giving the unsuspecting owner what appeared to be a nice patio. Now the owner is faced with the problem that the patio no longer provides a value-added element to the home. Something must be done. The owner could try to patch the cracks and pour a new corner where the old one broke away. However, the patches will be colored differently and so will the replacement corner. Rather than eliminate the problems, this solution will aggravate it.

Alternatives to the problem's solution. The contractor has to offer one or more solutions to the owner's problems. The contractor will likely probe the ground at the edge of the patio to determine if there is a perimeter footing. If there is, the concrete will be 10 to 12 in. deep. If there is only 3½ in. of concrete, there is no footing. Wire mesh might have been embedded in the concrete though. Since the corner of the patio is broken away, the contractor can determine if there is mesh by examining the edge at the break.

We discussed the need for planning expansion or control joints to permit the concrete to expand and contract. Patios usually do not have control joint and never have expansion joints. So a nonreinforced patio slab is a prime candidate for problems as it ages.

The contractor could offer to make repairs to the slab and then use an acid bath to strip off the aged surface and brighten the whole surface. However, the repairs will still show.

The contractor can offer to make the repairs, etch the surface, and apply a tile over the entire surface. It would only raise the surface about 5/8 in. and it would look very good. At some time though, the mortar/grout joints may crack because the slab is still subject to further deterioration.

The best alternative is to remove the old slab and replace it with a quality built one. The new one will have a footing reinforced with 2 rebars and a slab with 6 × 6 in. wire mesh. We will use a 3000- or 3500-lb concrete mixture versus the standard 2500-lb mixture. These two mixtures have much more cement and create a finer quality slab.

We can even add a pattern in the top with our mason tools. The patterns produce a wide variety of appearances. One is the checker board with 3- to 4-ft squares. Another is a series of parallel control joints that divides the slab into party areas and walks. A third is to make random curves that add a measure of informality to the slab. Finally, we can make a perimeter control joint about 12 in. in from the perimeter. These kinds of details are easily shown on a piece of note paper.

Statement of work and the planning effort. After serious discussions with the owner including the general costs of the recommendations, the owner elects to have the old damaged slab replaced with a quality built one. The first statement of work is *remove the old slab and replace it with a new one the same size in the same place.* We also must add *the height of the new slab will be 6 in. down from the exterior door step.* The quality needs to be stated: *The slab will be reinforced with a perimeter footing about 8 in. deep and 12 in. wide and it will be reinforced with two 1/2-in.-diameter rebars overlapped at the ends by 12 in. Furthermore, the concrete mixture will be 3500 lb per cubic yard.* The finish needs to be stated, too: *Using a control joint tool, the contractor will make a pattern in the fresh concrete according to the owner's desires.* Finally, we need to state the conditions for a turnkey job, so the *job is finished when all materials both old and excess are removed from the site and the landscape is restored. Any cement splashed on walls must be removed.*

The planning effort requires us to measure the dimensions of the old patio and determine how much of the perimeter requires forming. We need to assess how much the landscape has to be disturbed due to the setting of forms and working the concrete after it is poured. Since the slab is large, we will require two masons. The masons may be able to restore the landscape, but it would be better to give the work to a gardener.

Besides the work, we need to develop the contract and then plan a schedule of completion after the contract is signed.

Contract. The project we have outlined is very straightforward for a mason contractor. Thus there are no hidden aspects that could drive up costs. Therefore, we can offer the owner a fixed-price type of contract. Even allowing for the design pattern to be made in the surface, we are sure that this is the best alternative for both the owner and our company. The body of the contract would be something like this:

> We agree to remove and replace the old patio slab for the fixed price stated below. Included in the design and construction are a perimeter footing about 8 × 12 in. reinforced with rebars, a reinforced 6 × 6 wire mesh embedded in the concrete, and a pattern made in the concrete of the owner's choosing.
>
> In addition, we agree to remove all old and new waste materials, clean any spills or splashes from the walls, and restore the landscape to it's original condition.
>
> The terms of the contract are 20% at the time of contract signing and the balance upon completion of the work.

Material Assessment

Direct materials	Uses/purposes
Concrete	3500-lb Mixture
Rebars	Reinforce the footing

Wire mesh	Reinforce the slab
6-mil-thick Plastic	Cover the ground inside the forms

Indirect materials	Uses/purposes
2 × 8 Forms	Form the perimeter footing and slab
1 × 4 Stakes	Secure the form boards
8d Common nails	Used in building the forms
Plastic sheeting	Cover walls and spread over the ground

Support materials	Uses/purposes
Pneumatic jack	Break up the old slab
Small tractor or back hoe	Lift concrete into the truck for discarding
Mason tools	Masonry work
Power saw	Cut wood for forms
Transit	Used to set the height of the forms
Wheelbarrow	Transport concrete if necessary
Hoe and rake	Use to level concrete
Electric or gas trowel	Machine used to trowel the finish
Water hose	Washing and spraying if required
Shovels	Move concrete and gardening

Outside contractor support	Uses/purposes
Gardener	Restore the grounds

Activities Planning Chart

Activities	Time line (days) 1	2	3	4	5	6	7
1. Contract preparation	X						
2. Scheduling and materials	X						
3. Remove slab and form new one		X	X				
4. Pour and finish concrete				X			
5. Landscape					X		
6. Site cleanup					X		

Reconstruction

Contractor support. As we did in the first project, we must briefly examine the efforts of the contractor's office and staff personnel. This project requires similar activities, such as preparing the bid/contract offer, assembling the prices for materials and labor, allocating overhead for fixed and variable expenses, and making records of the job and associated details.

The contractor will require two masons, which may be himself or herself and one other. This means that the person is either an employee or must be contracted for from another contractor who regularly employs several masons. This may cause a slight increase in contract price, but the owner will not know it and really does not need to be concerned, since it is the responsibility of the contractor to pay the wages of workers.

The office person will likely set up the job after contract signing. This means that all material suppliers and workers will be notified about the acceptance of the contract and their need to be ready to go to work.

Materials and scheduling. Not a great deal of materials are required for the job. These can be picked up by the masons on their way to the job on day 1. The concrete will be ordered at the close of day 2, when we are sure that the forms, steel, and plastic are in place, for delivery early the third day. That same day someone needs to pick up the power darby/trowel and transport it to the site for smoothing and leveling the slab's surface. The machine will be returned to the rental store the next morning.

The gardener will either show up the afternoon of the pouring of the concrete or the next day to make a final assessment of materials needed to restore the lawns and gardens or pick up the expected replacement materials and arrive on the job as scheduled.

Remove slab and then form for the new one. The first part of the job is to break up the old slab into small pieces that can be loaded on a truck and hauled away. For this, we rent a pneumatic jack. We might lift the pieces in to the truck by hand or rent a back hoe with a bucket to aid in the job.

After the slab is out of the way, we rake the ground clear of everything and then lay out and build the forms for the slab. The series of tasks include the following; some are shown in Figure 7-4:

1. Set up the transit tool and set two to three 2 × 2 stakes at finished floor height.

2. Establish the lines for the form by taking references from the house and install batter boards and grade stakes. This is the time we verify accurate angles and dimensions.

3. Trench below the lines to a depth of about 8 in. and a width of about 15 in.

4. Cut one form piece to a length equal to the dimension of the slab from the house outward (for example, 16 ft.) and set it in place under the line.

5. Stake the piece at 3-ft intervals, making the top of the form even with the line for both height and position outside the line.

6. Repeat steps 4 and 5 for the form opposite the first one.

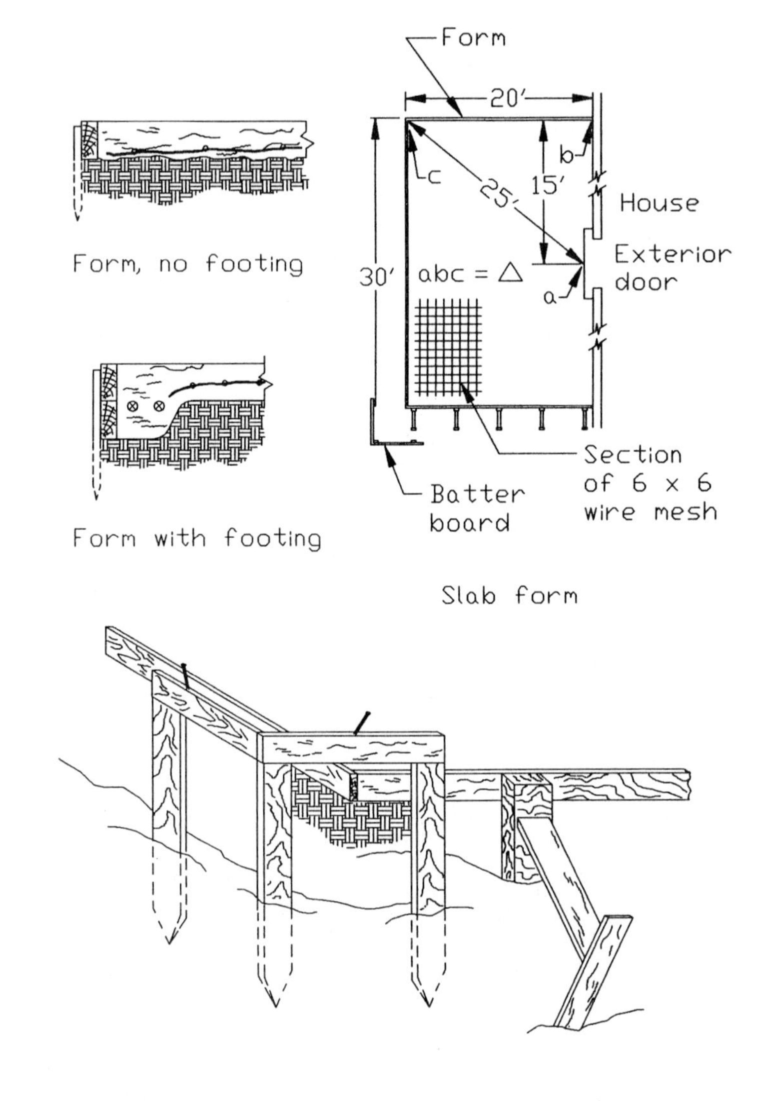

Figure 7-4 Forming for a new slab.

7. Install the end form by nailing it to the first and second piece installed and by bracing it securely.

8. Sculpture the ground inside the form to make an earthen inside form a minimum of 4 in. down from the form's top edge and to a depth equal to the form's bottom to create the footing. It needs to be about 12 in. wide. If the ground is very soft and needs packing, spray it with water and tamp the soil until packed. This eliminates settling and future cracks.

9. Lay in the plastic. If more than one piece is used, make sure that the lap is sufficient (12 in. or more).

10. Unroll and straighten the 6 × 6 wire. Using bolt cutters, cut several pieces to fit into the form. The overlap should be at least 6 in. We usually wire the overlap with wire wraps made from a roll of wire and lineman's pliers.

11. Two pieces of rebar need to be installed in the footing area. We can cut these to a length about 6 in. short of the full length, or we can bend the bar and let excess round the corner. Small stakes can hold the bars up off the bottom of the footing area, or we can lay them onto the wire and, while pouring the concrete, we can raise them into the concrete with hoes. It is better to have them placed before the concrete is poured.

12. Finally, we clear away all materials and other objects that can cause an accident while we are pouring and finishing the concrete.

Pouring and finishing the concrete. Pouring the concrete in the slab form is slightly different than pouring the sidewalk. Where we had two forms separated by 4 ft, it was easy to level off the concrete by screeding. In the slab, which may be 12 × 16 or larger, screeding is not very easy to do. In some cases a center board is installed temporarily until the leveling is approximate, and then it is pulled out and backfilled with concrete.

Another difference is in how we distribute the concrete. We usually chute the concrete toward the back of the form (the point farthest from the truck). Then we work forward to the front of the form. As we work forward, the driver either moves the truck or takes off chute extensions.

Another difference is how we work with the rebars and wire mesh. We generally reach through the concrete and pick up the wire so that it is not fully at the bottom but rather embedded in the concrete. While pouring the concrete in the footing area, we make sure that the rebars remain in their intended place.

We also make sure that the concrete is compacted against the form by using a vibrator (an optional piece of equipment to stir and vibrate the concrete).

After we have leveled and troweled the concrete's surface smooth, we will create the design pattern the owner wishes with our control joint hand tool. After this, we will take whatever precautions are necessary to protect the surface from weather and temperature.

Landscaping and cleanup. Before we begin to relandscape the disturbed grounds, we must remove all the forms. The customary way to do this is to first remove the nails from the braces and stakes holding the form and then we pry the forms up with a shovel. All forms are cleaned to remove the residue of concrete and cement and nails. All nails are pulled out of the stakes and braces. All the wood is loaded on the truck for return to the company site.

There is always spillage when pouring concrete, and there is always some leftover waste. Both of these harden just like any other concrete. So we are left with the job of removing this residue, and it is loaded onto the truck for disposal at a dump site. After the larger pieces are loaded on the truck, we rake and shovel up the loose stones and other evidence of concrete, wood scraps, and anything else, such as roots, stones, and the like, unearthed during the job and load these onto the truck as well.

At this point the gardener will take over and perform the relandscaping and restoration of lawn and gardens. The area surrounding the slab must be tilled, and new nutrients and mulch are added to the soil to neutralize the effects of the cement and concrete.

In the lawn area, the gardener sets out squares of sod and waters these heavily. In the garden area, the ground is covered with a weed-retarding cloth and plants are reset. Then the entire area is covered with decorative bark or something similar to match the remainder of the garden.

Concluding comments. The differences between replacing a sidewalk and replacing a slab patio are significant. The form is more extensive, there is reinforcement for the footing and concrete, and the finishing tasks are more difficult to perform. We also identified the need for a higher-quality concrete. Finally, we etched a pattern into the surface of the slab to add character.

PROJECT 3. RESTORATION OF DEFECTIVE SECTION OF A DRIVEWAY

Subcategories include forming; use of wire mesh; pouring and finishing the concrete; adding control and expansion joints; patching with the intent to cover the driveway with blacktop.

Primary Discussion with the Owner

Problem facing the owner. Over an extended time a section or two of the driveway has developed many random cracks and looks unsightly. It is also a major cause of frustration because of the unending need to weed and apply weed killer. The time has come to do something. The owner has heard about patching cement, which has an additive that makes it bond to old concrete. Some of the cracks are wide enough to force a bit of the cement mixture into, but many are not. The cracks are so random that it is impossible to cut a deep groove along the crack to make it possible to use the patching cement. There was some thought about breaking up the sections that are bad and carting away the pieces, but that is a major undertaking. The thought of covering the entire driveway with a blacktop from 5-gal cans was also considered and rejected.

The owner has called on the skill of a mason contractor, who supported most of the decisions that the owner considered, and went on to explain that the bad state of sections of driveway was caused by poor design and lack of proper reinforcement in the concrete. The contractor showed where there were too few control joints and no expansion joints. He or she also suspected that the original contractor saved a few dollars by not reinforcing the driveway with 6×6 mesh, or very likely, if wire was used, it was laid flat on the ground and not embedded into the concrete mixture. These two reasons account for most difficulty with driveways. A third problem occurs if very heavy trucks use the driveway. Their weight is so great that the poorly made driveway cannot support it and gives way.

Alternatives to the problem's solution. There are several alternatives to this problem. We should probably cut several control joints across the driveway in areas where there is no significant cracking problem. We may also be able to minimize the number of square feet of driveway to replace by making a control joint just outside the most damaged areas.

However, the most severely damaged areas must be replaced. In the replacement process, we will install expansion joints and control joints to largely reduce future problems.

Statement of work and the planning effort. The discussion with the owner resulted in the acceptance of most of our suggestions. These are translated into statements of work that include *cut four new control joints at the points indicated along the driveway and two more along the edges where the most damaged driveway exists.* We must also *remove and replace the severely damaged sections of the driveway. This accounts for 20 linear ft by 12 ft. The new concrete will be reinforced with 6 × 6 in. wire embedded in the mixture. Finally, we need to state that expansion joints will be used at each end of the new pour and a control joint will be made midway through the length.*

For the mason contractor, this is not a particularly difficult job. A lot of heavy work is required, but masons and laborers are accustomed to it. We must plan for a minimum of two workers on the job at all times. We must also rent equipment for a day to break up and cart away the old concrete. We must also plan to rent a control-joint-making machine and buy several blades. (Some mason contractors have these machines and depreciate their costs over time or as rental charges against jobs where they are required.)

We must coordinate with the concrete plant for delivery of the mixture, and we must have the forms in place before the truck and driver arrive.

Finally, we must schedule the job when the weather looks promising.

Contract. Once again we use the fixed-price contract type. We know how much work must be done, how many yards of concrete must be poured, how many control joints must be made, and how much expansion materials we need. We also know from experience how many days the workers will take to perform the work. Finally, we can obtain prices for all requirements. The body of the contract should state:

> We agree to remove and replace a 20-ft section of driveway, add reinforcement wire to the new section, separate the new section from the old with expansion joint materials, and add control joints where we agreed to with (Mr. or Ms. owner).
>
> The price includes materials and labor and site cleanup.
>
> Total Cost $×××.××
>
> This price is good for 45 days.

Material Assessment

Direct materials	Uses/purposes
Concrete	Replace driveway
Expansion materials	Create expansion joints
6 × 6 Wire mesh	Reinforce the concrete
Masonry blades	Cut control joints in concrete

Indirect materials	Uses/purposes
Form boards and stakes	Form the driveway
Nails (some double headed)	Holds forms and stakes in place

Support materials	Uses/purposes
Dump truck	Hauling waste concrete
Pneumatic jack	Break up old concrete
Power darby/float	Smooth concrete
Shovel, hoe, rake	Move concrete around in the form
Mason tools	Construction
Carpentry tools	Construction
Electrical cords	As required to operate the power tools
Gasoline	To operate the power tools
Water hose	For cleanup

Outside contractor support	Uses/purposes
None	

Activities Planning Chart

Activities	Time line (days)						
	1	2	3	4	5	6	7
1. Contract preparation	X						
2. Scheduling and materials	X						
3. Remove the old driveway		X					
4. Install the new driveway			X				
5. Make new control joints			X				

Reconstruction

Contractor support. Contractor support for this project will be very similar to the support needed for the replacement of the sidewalk that we discussed in project 1. The office personnel will need to compute the quantity of concrete needed, rent or schedule use of the power machines, schedule the workers, and coordinate with the concrete company for delivery of the concrete. Some of this will be done before contract acceptance and some after.

Materials and scheduling. Driveways are normally poured 6 in. deep. So when computing the amount needed, we simply multiply the length in feet by the width in feet by 1/2 ft. Then we divide by 27 to obtain the yards required. In this job our requirements are as follows:

Length	= 20.0 ft	$20 \times 12 \times 0.5 = 120$ cu. ft
Width	= 12.0 ft	
Thickness	= 0.5 ft	$120/27 = 4.44$ cu. yd (rounded to 5)

Wire mesh is sold in 6-ft rolls, so we need about 40 ft for the job. We also need to use 2-in. stock lumber for the forms or use metal forms.

Scheduling will be simple since we only need to pick good weather for the day we pour the concrete and make sure that the workers and concrete company get to the site on time.

Remove the old driveway. We described removing the old concrete in the study on replacing the patio slab. Before using the jackhammer, we must cut the control joints that establish the boundaries for the replacement. Then we can remove the old materials. If there was wire, it would be thrown away, too.

After the concrete is out of the way, we need to complete the task of making the cuts at each end of the area straight down so that we can set the expansion material in place against the old concrete.

All the old concrete must be taken to a landfill and discarded.

Install the new driveway. We set the forms in place first. There is one on each side of the driveway. We make them level with the old driveway and stake and brace them. Then we cut and lay in the wire mesh.

When the truck arrives with the concrete, we load the form with the concrete and level it with rakes and shovels. After a short while, we smooth it and, finally, rough it up with a broom or hand-held darby by dragging it across the 12-ft width. This adds traction.

Make new control joints. Part of our contract was to make a control joint in the new concrete and several in the old concrete. We use a power tool for this job. We install a masonry blade in the power tool, set the depth for 2 to 2½ in., and follow a chalkline across the driveway.

Blacktop the driveway. In the alternatives section of making restorations to the driveway, we mentioned that a driveway could be blacktopped to improve its appearance. We can cover up the old concrete with blacktop available in 5-gal cans or we can have a paving company apply a top as thin as 1 in.

In either alternative, the problems with cracks and the lack of control joints and expansion joints is *not* eliminated. However, these products can fill the cracks if they are minor or if they are up to 2 in. across. Each manufacturer of such coating supplies a small booklet of or pastes instructions on the can's.

Concluding comments. The work we have covered in this chapter is very heavy and strenuous. It requires people with strong backs and arms to move the concrete and prepare the forms, cut and install the wire and rebars, and the like. Compared to the study on replacing the sidewalk, this work is many times more demanding. Much larger areas are poured at one time, and the task of smoothing and leveling the concrete is much more physical. Usually, this work should be left to the skilled mason.

CHAPTER SUMMARY

In this chapter, we examined a wide variety of problems with concrete slabs, sidewalks, and driveways. We learned that the work is heavy and strenuous. Materials other than ready-mix concrete were needed to make the forms and add strength to the concrete. We learned the importance of control and expansion joints and where they should be placed. We also found out that concrete can be patched. More important we now know how to determine the quantity of concrete needed to fill a form, where to place the wire and rebars and how to finish the surface using one of several styles. Working with concrete usually requires two people for pouring a new driveway. However, one person can pour and finish a sidewalk. Finally, we must always remain conscious of possible physical stress and skin problems from the chemicals in the fresh concrete.

8

WOOD PORCHES WITH SCREENED PANELS AND WOOD COLUMNS

OBJECTIVES

To restore the functionality of the screened porch.
To understand the ways screen porches are built.
To determine which type of construction was used when the porch was built.
To be able to remove and replace the screen wire and moldings.
To be able to replace the screen door and closure.
To eliminate the cause of the rotting.
To restore wood columns.
To ensure that proper ventilation exists in the porch.
To repair the ceiling.
To replace the ceiling light fixture.

OPENING COMMENTS

Conditions or circumstances that require corrective actions. Screened porches provide a great deal of pleasure for the owner and their family. Figure 8-1 is an illustration of one made from panels. They are usually very serviceable year after

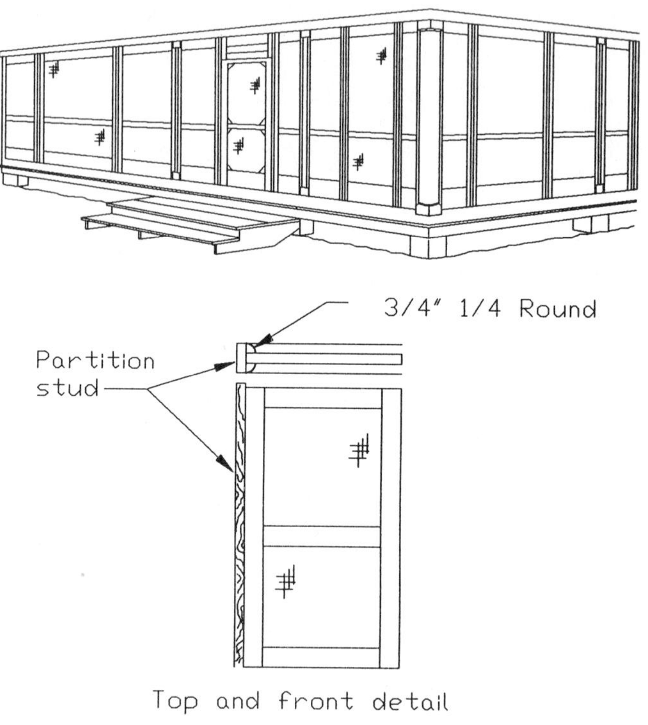

Figure 8-1 Screened porch.

year, unless normal living activities create problems. The list could include a hundred causes for damage and wear and tear. Some of the more common ones are torn screen wire from children's toys and falling objects and furniture, broken screen doors that were not very strong to begin with, rotted frames where rain and snow were permitted to seep into the wood, mildew where air was blocked from circulating, columns that rotted or separated, poor upkeep, lack of proper painting, damaged ceilings, inoperative ceiling fans and light fixtures, and so on.

Contractor responsibility. The contractor who is called to the job site needs to be well experienced in finish work. (There are rough carpenters who erect buildings and there are finish carpenters who install the trim and cabinets, build cornices, and do other jobs where wood pieces must be accurately fit.) Screening in a porch requires the skill of a carpenter with some cabinetmaking experience. The need arises when the porch is enclosed with panels versus 2 × 4 construction. Therefore, the first responsibility is to examine the type of construction used and make this known to the owner.

If the panels are unit construction as Figure 8-1 shows and there is damage to the wood, cabinetmaking skills will be required to make a new panel. If the panels must be removed and replaced, then molding must be taken off on one side and later replaced. The contractor must explain about the mortise-and-tenon joints used in its construction. In all likelihood, the panel will be made in the shop versus on site.

If the screen porch was framed with standard 2 × 4's or 2 × 3's, the contractor needs to show this to the owner. The explanations on replacing defective members pertain to taking up old materials and replacing them. Screen wire, moldings, and the like will be torn off and replaced on site with new materials. All construction will be on site.

If there is damage to the columns holding up the roof, the contractor needs to show the owner that the columns are one-piece or built-up types. Then he or she can suggest the most suitable replacement action. One-piece columns can usually be replaced by raising the roof slightly, slipping out the old column, and then slipping in the new one. But to do this we must remove the wood frame attached to the post and the screen wire attached to the framing.

If, on the other hand, the column is built up with a double 2 × 4 core and 1 × 6 exterior, repairs can be made without removing the post unless the 2 × 4's have rotted as well. Even in this situation the owner must know that the screen panel or frame attached to the post must be removed before work begins to restore the post.

If there is mildew, we can clear that up quickly, but we also have to eliminate the cause. In this effort, we explain that poor circulation and moisture combine to provide a perfect home for mildew growth. We need to find a way to increase circulation. In this same environment, we will find corrosion. Lamps and ceiling fans are the most frequent items to show corrosion. Hinges, door closures, and locks may also exhibit corrosion. Most of the time a good paint system will stop this problem, but sometimes the units must be replaced.

Homeowner's expectations. All this detail may be essential for the contractor, and it is important for the owner to know about, but the owner needs to have a quick evaluation and a fair price for the work that must be done. If the wire is torn, he or she wants it fixed quickly and efficiently with minimal cost.

The owner also expects that the required work is surface or visible and does not involve major problems. Major problem include rot and decay of either the columns or screen panels or frame or on all the components of the screen porch. The owner who finds out that rot is present should expect that replacement costs will be expensive. For some homeowners, some insurance policies will cover all but the first $250.

If the work requires the needs of a cabinetmaker to manufacture one or more new panels, off-site work will be required and this may increase the duration of the job, as well as the overall costs.

After discussions with the contractor, the owner may feel confident that he or she can do some or all of the job himself. This would be a good time to elect this option. It is fair to the contractor and eliminates his or her time and effort in composing a bid/contract that may not be exercised. On the other hand, the contractor should prepare a bid/contract for the portion of work not being done by the owner.

Scope or types of projects to solve. We will focus on several parts of the screen porch in the projects. The first will be the simplest, removing and replacing damaged or rotted screen wire. The next will be the replacement of the screen door and hardware. The third will be the removal and replacement of damaged 2 × 4 (or 2 × 3) framing to which the wire is stapled. The fourth project will be to replace a damaged screen panel (this study includes details about the joinery employed in the construction). The fifth project will be to restore the ceiling and attached electrical devices. In the sixth we restore a wood column.

PROJECT 1. REPLACE ROTTED OR DAMAGED SCREEN WIRE

The subcategories include removing the screen molding and replacing it after the wire is changed; removing and replacing the screen wire or cloth; stretching the wire as it is installed; priming the new molding and painting the top coat.

Primary Discussion with the Owner

Problems facing the owner. The owner of the screen porch is faced with damaged and/or rotted screen wire on one or more panels. If he or she elects to replace the material with new wire, the job will require removing the molding covering the staples, removing the wire, and replacing it and the molding.

If the porch is made with panels as shown in Figure 8-1, there are separate pieces of wire on each lower and upper area. If the damage is on the lower one, then only that one needs replacing. If the damage is from weathering and all panels need the wire replaced, then four pieces of molding must be removed from each panel and replaced after the wire is replaced. All of a sudden the job has escalated to much more than a weekend job. The molding normally used is called *screen molding.* It is about 3/4 in. wide by 1/4 in. thick and usually has an imprinted design. When these pieces have been in place for a long time, they are extremely brittle and usually crack or split when pried off. They also have a nasty habit of breaking at the place where the nail is.

If the porch is framed with 2 × 4's over which the screen wire has been stapled, then the staples are usually covered with 1⅝-in. by 1/4-in.-thick lattice. This makes a pleasant appearance on the outside. Even though there might be a chair-rail-height 2 × 4 across the panel, the wire is usually a single piece from head to floor. Thus, when either the upper or lower wire has been ruptured, the entire full-length piece must be replaced. Long pieces of lattice along the bottom and top will be removed even though the panel being repaired is only 36 in. wide. This concept is shown in Figure 8-2. Lattice splits as easily as molding. So there may be some rather long pieces to replace.

In the first situation, smaller pieces of wire are used, and these may be easier for the owner to replace. In the second situation a piece of wire about 8 ft long is installed. It may be a problem for the owner to install and get it taut. A qualified carpenter would have no difficulty with either situation.

Another problem facing the owner is the decision regarding matching the type and color of screen wire. Old, weathered, aluminum and copper screen wire self-protects with oxides. This causes the aluminum to turn blackish and copper to turn greenish. The oxides extend the life of the wire. When new aluminum or copper wire

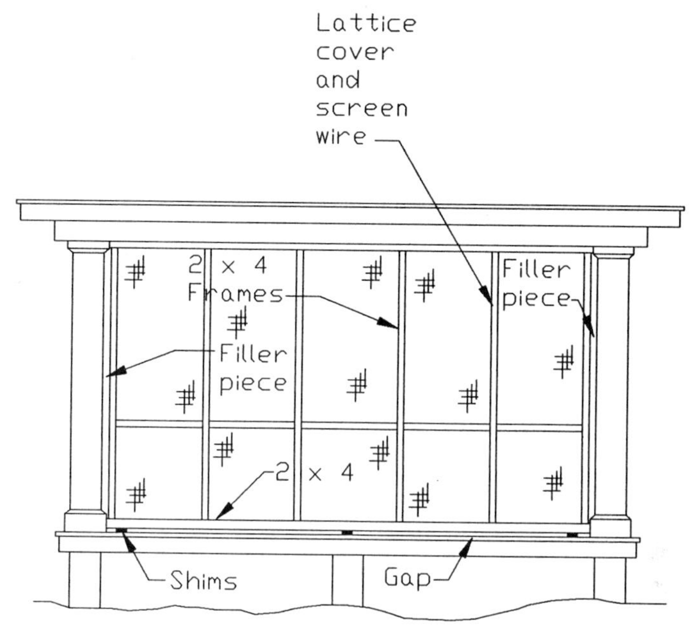

Figure 8-2 2 × 4 Framed porch.

is installed, at least one season must pass before the oxidation is complete and the color matches the old wire.

When nylon cloth wire is used, it is usually black to begin with. It almost never changes color, and it never oxidizes. However, mildew will accumulate.

Finally, the owner will have to work from a ladder part of the time, and this may pose a problem, since the wire must be stretched with one hand while the other hand operates the staple gun.

Alternatives to the problem's solution. There are very few alternatives to the problem. They include the following:

1. Remove and replace all the wire so that we are not faced with replacing a piece this year and another next year, and so on.

2. Use the same wire as before: copper to replace copper, and so on.

3. Use a wire that can take knocks without ripping.

4. Use a wire that is overall cheaper.

5. Use the expertise of a carpenter or do it yourself.

Statement of work and the planning effort. For this project we expect to replace all the screen on the porch with the work performed by a carpenter. The statements of work begin with *remove and replace all the screen wire on the porch,* *including the wire in the door.* Since the price varies greatly between copper and aluminum screen and nylon cloth, we better specify the kind: *all new wire will be aluminum.* We should also specify that *all molding and lattice strips that are defective in any way will be replaced with new materials.* Finally, we need to specify that *the wood frames and moldings will be painted with two coats of exterior latex of the owner's choice of color.*

The planning effort for the contractor requires taking measurements of the width of the screens and calculating the length, determining the number of lineal feet of new molding and lattice, estimating the amount of paint, and planning how many days it will take to complete the job. The plan must also include two trades, a carpenter and a painter. The painter in this study is a subcontractor. Thus the painter needs to provide a cost estimate and a time when he or she can fit in the job.

Contract. The contract for this job will be a fixed-price type. Even though the painter will submit a bid, we will only provide a single price for the entire job. The body of the contract should state:

> For the fixed price shown below, we agree to replace all the screen wire on the porch with aluminum wire, use new molding and lattice where it is required, and apply two coats of exterior paint to all the wood on the porch panels and door. The floor and ceiling are exempt, but columns will be treated as part of the frame of the screen porch.
>
> Total cost $×××.××

Material Assessment

Direct materials	Uses/purposes
Screen wire	Replace old worn wire
Staples	Staple wire to frames
Screen molding and lattice	Cover staples and add a finish
Latex exterior paint	Paint the wood parts of the porch, screen panels, and door
3/4-in. Brads	Nail the molding
3d or 4d Finish nails	Nail the lattice

Indirect materials	Uses/purposes
Drop cloths	Cover the porch floor to prevent paint spray

Support materials	Uses/purposes
Carpentry tools	Construction
Step ladder	To reach the heads of the panels
Sawhorses and plank	Workbench to cut molding

Outside contractor support	Uses/purposes
Painter	Paint the wood

Activities Planning Chart

Activities	Time line (days) 1	2	3	4	5	6	7
1. Contract preparation	X						
2. Scheduling and materials	X						
3. Remove and replace the screen		X	X				
4. Paint the wood				X	X		

Reconstruction

Contractor support. The office personnel will estimate the quantities of materials, apply overhead costs, and solicit a bid from the painter. These items will be included in a worksheet for internal accounting and form the basis of the bid/contract offer.

Materials and scheduling. All the materials for this job can easily be obtained from a variety of builder's supplyhouses. So they can be obtained on the day work begins or prior to the start of the job.

Scheduling must take into consideration the weather and availability of the carpenter and painter. With respect to the carpenter, he or she will be expected to take several days to complete all the work. On an average, for each full-length screen panel, regardless of width, it will take about 1/2 hour to remove the molding and old wire, and about 1½ to 2 hours to install the new wire and either replace the old molding or cut, fit, and install new molding. This translates to about six to eight panels per day. If the panels are made using the cabinetmaking technique, the time for replacing the same amount of wire will take slightly longer since there is more molding to remove and reinstall.

The painter should apply a prime coat to all new moldings before they are cut and installed. However, it may not be practical to do this. In that event the painter will schedule a day to prime the new wood and also in that same day begin to apply the top or finish coats. Many exterior latex paints dry within hours. Due to the requirement for quality, exacting workmanship, the painting takes time to do. The painter must cut in all the edges of the molding without getting paint on the screen wire. It, therefore, will probably take an hour or more per screen panel for every coat.

Remove and replace the screen wire. We have examined much of the work in our earlier discussions. Here we concentrate on the actual task performance. First, the moldings must be removed. We do this by using a chisel as a wedge to pry up the molding. Once the chisel is driven under the edge of the molding, we push down and then twist it to pry up the molding. If we do this where the brads are

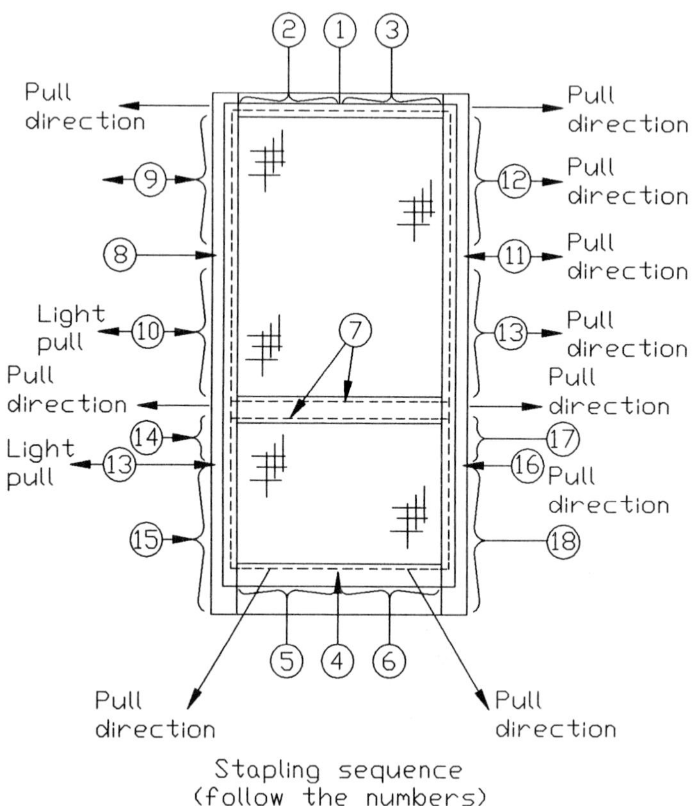

Figure 8-3 Installing screen wire.

installed, we reduce the chances for cracking or splitting the wood. After freeing a piece we *never* drive the brads back out through the front of the molding; we *always* pull them through from the back with a pair of linemens' pliers.

With the molding and all the brads removed, we strip the old wire from the frame and remove all the brads and staples. Then we are ready to install the new wire. Figure 8-3 shows the sequence we use. The steps are as follows:

1. Cut a length of screen wire about 3 to 4 in. longer than the full length of the opening.

2. Center the wire from side to side along the top.

3. Staple the screen wire in the center at the top about 1/4 in. in from the edge.

4. Install one staple at the bottom, temporarily, to make sure that the wire is centered.

5. Then follow the stapling patterns shown in Figure 8-3. While stretching the wire with one hand, we staple with the other.

a. Complete patterns 1 and 2 across the top.

b. Complete patterns two and three along the top. These two require both a sideway and downward pull on the wire while stapling.

c. Install one staple in the middle of the long side and another opposite it; pull the wire tight. This creates patterns 5 and 6 and 7 and 8. Complete the patterns as numbered.

With the wire stapled, we reinstall the molding with brads and set each slightly. Then we use either a sharp 3/4-in. wood chisel or sharp utility knife to trim away the excess wire along the outside of the wood molding. The job is done.

Painting. We are not going into the types of paint to use since there are several earlier discussions in other chapters on this subject. The painter must fill the brad holes before completing the job.

Concluding comments. The job of removing and replacing screen wire is exacting but certainly not very difficult. If the owner tries to do the job himself or herself and makes several mistakes before doing the work well, the cost of wasted materials is not all that great.

The main points to remember are to try to save the molding if possible, remove all brads from the molding from the backside, install the wire tightly and straight, and set all brads as they are nailed in. All new molding should be primed before it is cut, fitted, and installed. This simplifies the work. Finally, we must not get paint on the screen wire.

PROJECT 2. REPLACEMENT OF THE SCREEN DOOR AND HARDWARE

The subcategories include fitting and hanging the new door (wood and metal); installing a new lock unit; installing the door control assembly; fitting the stops; painting the wood door.

Primary Discussion with the Owner

Problem facing the owner. In a screened-in porch there is always a wood screen door that leads from the porch to the backyard, garden, pool, or garage area. Over time, this door takes a lot of abuse. Generally, it is hung with either spring-loaded hinges, has a screen door spring, or has a screen door closure. Each of these devises forces the door closed. Often the force is much too great and the door bangs. Since these wood doors are only $1\frac{1}{4}$ in. thick, they cannot take the abuse for more than several years before the joints part or splits occur. Then the owner is faced with the problem of either making repairs or replacing the door. In this section we will develop the techniques to replace the old door with a new one.

Alternatives to the problem's solution. There are several alternatives worth examining. The door can be purchased from a variety of outlets. When this option is selected, there is a strong likelihood that the middle rail in the door will *not* match the middle rail in the screen wire panels. If it did not before, then nothing will have changed. If the old door had a matching-height middle rail, then a new one would need to match as well.

Many times a custom-made door is built by the contractor who built the screen panels. This ensures that the horizontal line of the rails is unbroken. When replacing the door we would not be able to use a standard off-the-shelf door, but would need a new one built by a cabinet shop.

So, our alternative on this aspect of the project is to match the old door and make sure its design does not interrupt the lines of the porch screen panels.

Next we need to select the type of hinges to use on the new door. We can use the butterfly type if the old door used them. Butterfly hinges have built-in springs. Or we can use a loose-pin hinge and mortise it into the door and frame. If the old door used this type, we should also use it, but we may have to fill the screw holes to ensure a solid anchor for each half of the hinge. If we use the loose-pin hinge, we will need a door closure or spring to provide automatic door closing.

The old door may have been equipped with a screen door spring. The owner should tell us if he or she wants this type of closure or the type that has a plunger, which prevents banging and limits opening as well.

The last alternative is the type of door locking set. We might be able to use the old one, but if it was used for a long time, the mechanisms are usually worn. We should recommend a new unit. There are several types. Most are equipped with a slide control that prevents the handle from moving. We should suggest one with a locking device.

Statement of work and the planning effort. The first requirement is to select a door that meets the requirements of the design: *The screen door design must conform to the design of the porch screen panels in that the rails are in alignment.* Next, we need to specify the hardware, such as *all hardware shall be replaced. The same type of hinges will be used, and an automatic door closure will be used to control the movement of the door.* We need to make a provision in case the standard door cannot meet the needs of the first statement of work: *The quality of workmanship employed in the manufacture of the door, if a standard one cannot be purchased, must include mortise-and-tenon joints and knot-free fir and be glued and clamped. The styles and rails will match the old door for all dimensions.* The door will need installation and later painting; the statement of work reads *install the door and hardware using quality fitting techniques and a quality exterior latex painting system.*

The planning effort requires definition of the best alternatives that meet the specifications. Acquisition of the door is the most critical and it could take a day or several weeks. Estimating the job will be simplified once the decision is made to use a standard or custom-made door. The installation is a 1-day job for one carpenter. The painter will need 2 days to paint the door, even though the work requires only about 2 hours per coat.

Contract. In all likelihood, a fixed-price contract type should be used for this job. Even if the door needs to be made, the price should be known within several hours after contacting the cabinet shop. Standard rates for painting can be used, as well as carpenter labor. The body of the contract could state:

> We agree to replace the screen door with one that matches the old one for style, quality, and type of hardware, with the addition of an automatic closure. The door will be primed and painted with an exterior latex painting system.

Material Assessment

Direct materials	Uses/purposes
Screen door	Replace the old door
Hinges	Replace the old ones
Door lock set	Replace the old one
Automatic door closure	Added to the door
Latex paint system	Prime and two finish coats
Sandpaper	Smoothing

Indirect materials	Uses/purposes
Stops	Possible replacement
Wood filler	Fill old holes if required

Support materials	Uses/purposes
Carpentry tools	Construction
Painting tools	Apply paint system
Ladder	Painting
Sawhorses	Work surface to prepare the door
Power saw and cord	Eases installation
Power drill and cord	Simplifies installation

Outside contractor support	Uses/purposes
Painting contractor	Paint the door and trim (if required)
Cabinet shop	Build a custom-made door

Activities Planning Chart

Activities	Time line (days) 1	2	3	4	5	6	7
1. Contract preparation	X						
2. Scheduling and materials	X						
3. Remove and replace door		X					
4. Apply painting system			X	X			

Reconstruction

Contractor support. The contractor's office staff, estimator, or manager may have to prepare the bid/contract as well as locate a suitable door. This may require travel to several builder's supply houses and a cabinet shop. Certainly, there will be follow-up time needed for obtaining the bid to manufacture a custom-made door. A call to the painter is all that is required to obtain a price for his or her work. This should be followed up with a written bid from the painter. The final bid/contract may take several days to a week to complete.

Materials and scheduling. We have already established the variety of materials required to complete the job. The carpenter assigned to the job will pick up all the materials on the day of the job. Likewise, the painter will bring the paints and tools needed with him or her.

Scheduling hinges on the expected date, after contract signing, that the door will be available and the carpenter can be freed from another job or completes another job. Since this contract is a very small one, the contractor will try to fit it into a slot between jobs or when there is a lull between phases of a larger job. The painter will do likewise.

Unless there are extenuating circumstances, the workers can perform the work without anyone present at the house.

Removing and replacing the screen door. The tasks are very routine for the craftsmen.

1. Remove the old door by freeing the spring (if there is one) and taking out the screws that hold the hinges in place.

2. Remove the old striking plate from the front of the door jamb.

3. Cut the new door for the proper height.

4. Fit the door into the opening and mark for proper (1/8 in.) space allowance between door and jamb.

5. Mark the door for the placement of the hinges and install these onto the door. If mortise hinges (loose pin) are used, the placement needs to match the ones in the frame.

6. Position the door and install the screws in the frame.

7. Install the new lock set and its striking plate.

8. Install the new automatic closure.

9. Test the door for travel and closing. Make adjustments as needed.

The job is done.

Painting system. Refer to other sections where these data are provided.

Concluding comments. The job of replacing the screen door requires skill of the carpenter and a certain amount of painting skill as well. Of the two parts, the owner can likely do the painting. Cutting and fitting a door looks easy, but is not. Many things can go wrong, especially in the fitting and hanging tasks.

The job can be modest in price to very expensive. A wide variety of standard doors are made to the size the owner will need. The contractor should provide a quality one, but if money is a serious consideration, a much cheaper one will have to be substituted. If the door must be custom made to specification, the price will always be greater than for a standard door.

PROJECT 3. REMOVE AND REPLACE DEFECTIVE 2 × 4 FRAME MEMBERS

Subcategories include removing the molding and old wire or screen cloth; taking out the rotted vertical and horizontal members; preparing and installing new treated members or naturally resistant wood members; repainting the new wood; installing the new wire; nailing the molding over the staples.

Primary Discussion with the Owner

Problems facing the owner. This project can be very serious and can cost a considerable amount depending on the severity of the damaged materials. In the construction of a screened porch made from common 2 × 4's or 2 × 3's, one piece must be placed horizontally along the floor. Those built with good design principles never nail the sole plate (base) pieces to the floor. Rather, they insert a 1/8- to 1/4-in. shim between the 2 × 4 and floor. This slight space provides an avenue of escape for rain water. All porches are sloped away from the house at least 1/4 in. per 10 ft of run. Figure 8-4 shows a close-up of a base piece flat on the floor and another with shims installed.

If the porch has rotted sole plates, the owner faces the removal of these pieces and the vertical members in some cases. This means that the screen wire must be removed and discarded. Moldings nailed across the bottom will be thrown away. The rot may also extend to the bottoms of the 2 × 4's. If whole sections such as the south wall have to be removed, the work requires removing the head pieces as well. These are the top horizontal members that are nailed to the soffit.

Since all screen wire and moldings are removed to get to the 2 × 4 frame, we face the cost of replacing them. Old wire still serviceable can not be reused, primarily because it was trimmed very closely when installed.

Alternatives to the problem's solution. One of the most important decisions to make when having the restoration work done is to eliminate the need to replace the members again for the same reason. To this end, we can use treated materials and ensure that water can readily escape from the porch under the 2 × 4.

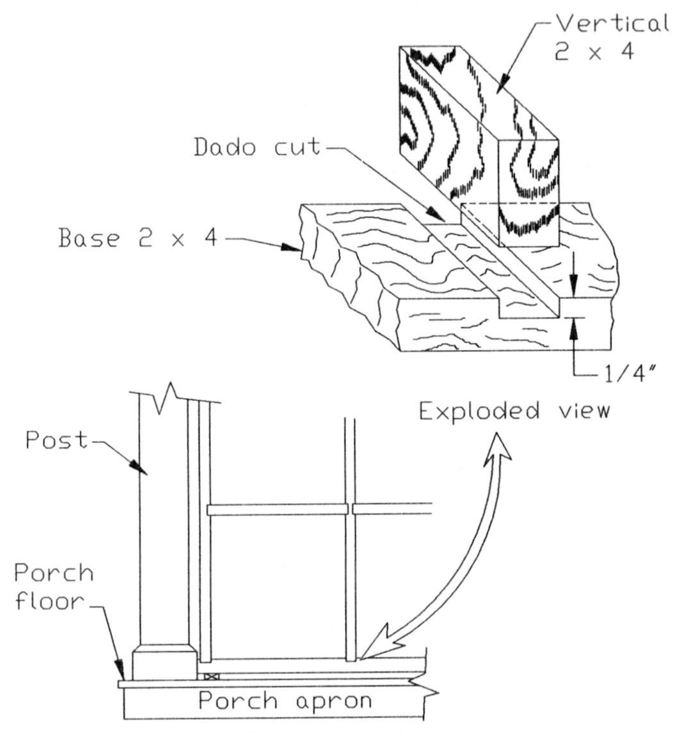

Figure 8-4 Close-up of the base 2 × 4 with water gap.

We must plan to sacrifice a lattice strip on the inside and outside of the base. In these pieces we cut inverted V's. They act as weep holes and allow the water to escape. Ultimately, these will trap some water and eventually will rot. But their cost is very nominal. The main frame is sound.

We can examine the vertical members for damage and replace only those that have physical damage. This will reduce the costs some. Removing the rail pieces will likely split them, so we better plan to replace them with new materials. New wire and some new moldings will be required and, finally, a new paint job will be needed.

Statement of work and the planning effort. Various parts of the project need to have specific statements of work, beginning with *replace all damaged or rotted members of the frame with treated lumber.* Since the wire will be stripped away, we must *replace all removed wire with new wire of the same kind, for example, copper for copper.* Also, *replace all wood moldings and lattice that cannot be saved.* In the area of improvement, *ensure that a 1/8-in. gap is built between the porch floor and bottom of the sole base member to permit water to escape from the porch.* We also need to state that *when applying the new painting system, no paint must touch the wire beyond 1/16 in.*

This project is the most intensive one in this series. We will literally be tearing the screen panels apart to get to the frames. Our expected labor costs will be much greater than for the earlier projects discussed. We can expect one carpenter to do the work. Or we can employ two carpenters if we want to complete the job in less time.

We need to understand that nothing can be precut or premanufactured. So all work must be custom fitted on site.

Several days are required for completing a section of wall about 20 ft long. A single section of wall of about 10 to 12 ft can be restored in an average of 2 days. If the section of wall includes the door as well and we must rebuild the jamb, then added time needs to be included. Then we must allow several days for painting.

Our planning will also take into account the weather and when it would be convenient for the owner to have us do the work.

Contract. The contract can be the fixed-price type since we can accurately measure the work to be done. The body of the contract should address the statements of work. We can assume that the contract requirements from projects 1 and 2 in this chapter apply here as well. If true, all three parts should be included. For this project, we will consider that the wall has no door. Therefore, our bid will encompass the replacement of the framing and the contract statements pertaining to wire replacement and painting. For example:

> The price stipulated below applies to the following work and is effective for 45 days from the contract acceptance.
>
> We agree to replace all defective or damaged framing of the screened porch with treated materials. We further agree to improve the design to permit water to readily escape from the porch. All new framing materials will be primed and painted after a 30-day drying period.
>
> We agree to replace all the screen wire on the porch with aluminum wire, use new molding and lattice where it is required, and apply two coats of exterior paint to all the wood on the porch panels and door. The floor and ceiling are exempt, but columns will be treated as part of the frame of the screen porch.
>
> Materials and labor $×××.××

Material Assessment

Direct materials	Uses/purposes
Treated 2 × 4's	Framing members
10d Galvanized casing nails	Nailing members
Treated shims	Cut from 2 × 4's
Molding and lattice	Cover wire at staples
Brads	Nail molding
Screen wire	Screening
Staples	Fasten wire to frames
Latex paint	Paint wood

Indirect materials	Uses/purposes
Water	Cleaning painting tools
Roller and pad	Disposable painting materials

Support materials	Uses/purposes
Carpentry tools	Construction
Painting tools	Painting
Ladder	Working surface
Sawhorses and planks	Working surface or bench
Power tools and cords	Ease work

Outside contractor support	Uses/purposes
Painter	Paint the wood

Activities Planning Chart

Activities	Time line (days) 1	2	3	4	5	6	7
1. Contract preparation	×						
2. Scheduling and materials	×						
3. Remove and replace framing		×	×	×			
4. Rescreen frames			×	×			
5. Painting system					×	×	

Reconstruction

Contractor support. The office personnel, manager, or estimator will prepare the contract for this job. He or she will accumulate prices for the new materials, contact the painter for a price, and determine how many days labor will be required. Even though the work involves using 2 × 4's, we know that the work is classified as *finish* type. All joints and miters must be as perfect as possible, since everything is visible. Another reason quality workmanship is required is the need to make the porch insect free. The pieces fitted to the walls and posts or columns must be accurately fit.

The estimator uses this knowledge to plan how many days for labor. He or she also knows that the job can be either a one- or two-person job. From the activity list, we see a total of 3 days are needed for our project for one carpenter. If two are sent to the job, we can expect to complete the work in under 2 days, but we will have to charge for two full days (four worker-days versus three). This will drive up the costs.

The contractor will also require the carpenter to pick up the materials on the first day of work, so we need to add approximately 1 hour for this.

Materials and scheduling. The materials shown in the materials assessment can readily fit into a pickup truck, so the carpenter will have no difficulty transporting them to the job. There are not that many tools, machines, and support materials to carry, so these can also fit into the truck.

The carpenter must be assigned to the job after the contract is signed. He or she must be told that the job needs to be done in 3 days, weather permitting.

The painter must be notified of the date when the carpenter is finished so that he or she can determine the appropriate time to paint the new wood. Remember, we must allow 3 to 4 weeks for the treated lumber to dry out.

Remove and replace the frame. On the first day at the job, the carpenter will remove all the wire from the frames to be replaced. All moldings that can be saved will have the nails removed and the molding will be stacked out of the way. It is possible that some of the frames will also be removed and discarded.

New frames will be installed a piece at a time, beginning on the second day, and probably most of them will be in place by the end of the day. However, the carpenter may elect to complete a section of porch screen wall with wire and molding before going onto the next section. This option neither costs more nor saves time; it is just a personal preference.

Along with installing the frames, the carpenter must cut to fit special pieces between the wall siding or brick and 2 × 4 frame or between the post and 2 × 4 frame. Figure 8-5 shows a sectional view of each. Although these look difficult to cut, they are not when done by a skilled carpenter.

The carpenter will use nailing techniques that minimize the nails' visibility. This adds quality to the job. Toenailing will be used. Many of the nails will be driven in places where they will never be seen. Others will be set and later filled by the painter.

The screen panels will be completed as described in project 1.

Painting system. The exterior latex semigloss painting system will be used. A prime coat and two finish coats will be applied. The color will be the owner's choice.

Concluding comments. This type of screened porch construction can look either very professional or shoddy; 2 × 4's can be joined accurately or can fit poorly; nails can cause splits and cracks and show, or they can be strategically located and installed where they do not damage visible surfaces. The special pieces can fit snugly and look very good, or they can fit so badly that they might as well not be there.

The 2 × 4's are seen from the inside of the porch. The workmanship must be high quality. The materials must be very good quality, not utility grade. The outside must also be constructed with quality workmanship and materials. Screen moldings and lattice joints must fit snugly and accurately.

The painting system must be proper and the workmanship must be first rate. No paint must be applied to the screen wire or to bricks or siding. Accuracy is extremely important.

The owner could do this project if the quality standards are strictly adhered to. The owner, if only semiskilled, may have waste because of the recutting required for quality fits and joints. Likewise, the owner would have to have good painting skills to do a quality versus a sloppy job of painting.

Carpenters and a painter, on the other hand, would not create much waste and would add the quality to the project.

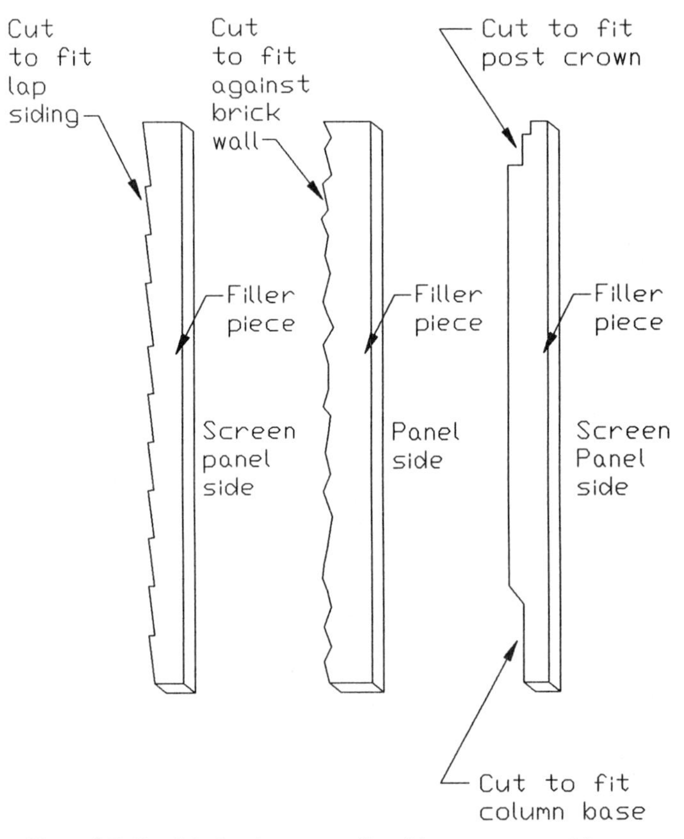

Figure 8-5 Special piece between wall and frame or post and frame.

PROJECT 4. REMOVE AND REPLACE A DAMAGED SCREEN PANEL

Subcategories include removing the quarter-round molding, lattice, or base boards; removing the old damaged panel; repairing or manufacturing a new panel; prepainting the wood; application of new wire and molding; the reinstallation of the panel.

Primary Discussion with the Owner

Problems facing the owner. A screen panel, as we will understand shortly, is a single unit installed between columns or from wall to column. It is a self-contained assembly as shown in Figure 8-6. It is made using cabinetmaking joinery. Rails and stiles are joined with either double dowels or a mortise-and-tenon joint and exterior glue. As we see in Figure 8-6, the various parts are designed to provide design and shape to the porch and to form the basis for stapling the wire. The screen molding is nailed over the wire.

Owners can have several problems with screen panels. First, the screen wire can rot or be torn and need replacing. Project 1 covered this problem. On a more serious

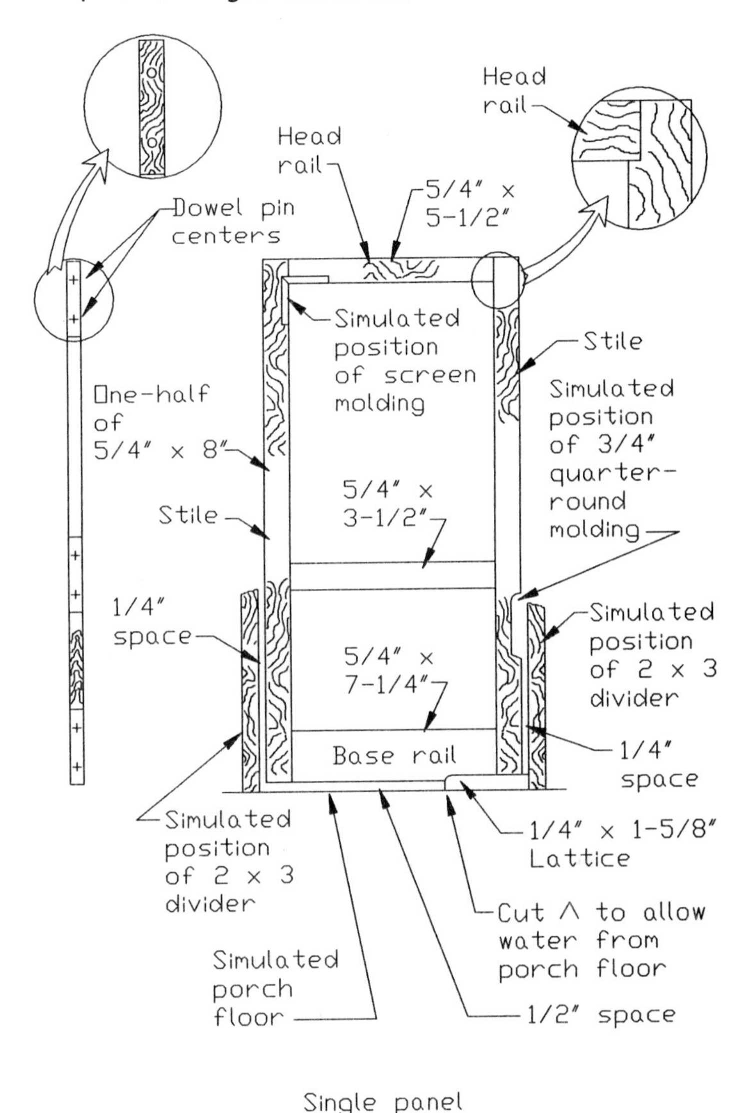

Figure 8-6 Panel screen unit.

note, the wood used in the construction can rot due to improper painting or wood that contained decay during construction, or perhaps a physical blow broke a stile, rail, or joint. Any of these are serious enough problems to call a qualified contractor or cabinetmaker.

Alternatives to the problem's solution. We will not discuss the problem of screen wire replacement, since that has been covered before. Rather, we will examine alternatives where the wood in the panel is at a point that it needs replacement.

First, we will not be able to make repairs while the panel is installed in the porch. Our only alternative is to remove it from the wall and take it to a shop where there are proper tools to make the repair.

Next, we will have to assess the damage and determine how best to make the repairs.

1. Should we replace the damaged stile or rail? Can we do this if the damage is to an interior member of the panel? Can we do this if the damaged member is a rail?

2. Should we separate the stile from the rails attached with glue and dowels or mortise-and-tenon (M&T) joints? This could be a real problem.

3. Should we build an entire new panel? At what point should we advise the owner that the best alternative is to build a new panel?

When we explore option 1, we are faced with the problem of trying to replace an interior member, but the only joint we can use is the butt type. There is no way to make either the double dowel or M&T joint. Therefore, the repair job will be less durable than the original design. If the damaged rail is the head rail or bottom rail, we are again faced with the problem of making a dowel or M&T joint, and again we cannot use any but a butt joint.

There is one alternative we can add to the butt joint to add strength. It is a flat metal plate about 1 in. wide by 4 in. long. After gluing the new piece in place, we route a groove in the face of the panel across the joint as Figure 8-7 shows. We make the groove about 1/4 to 5/16 in. deep. The metal plate is 1/8 in. thick. After it is installed, we fill the rest of the groove with a wood putty made from powder and mixed with water. After it dries, we sand it smooth and it is ready for painting.

When we explore option 2, we remove the screen wire and try to separate the stile from the rails at every point—top, bottom, and middle rail. If the glue joint has deteriorated, it is possible that the pieces will separate. If not, our solution is to cut the stile free. We do this by cutting the rails at the point where they contact the stile. Granted, we lose about 1/8 in. of rail, but that is not important. This technique permits us to replace the defective rail or stile. We then use the double-dowel technique to join the pieces back together. No metal plates are needed.

When we explore the third option, the damage must be extensive. Somewhere near one-third of the panel needs replacing. There must be more than three joints affected, and more than one general area must be damaged. Furthermore, we expect to see surface damages such as gouges or splits.

We really find that the time expended to separate the entire frame and cut customized pieces to fit into the panel exceeds the time and effort it will take to build a new frame from scratch. The trade-off is a reduction in labor and some added cost in materials, but the quality of the new panel will equal or exceed that of the old unit.

Statement of work and the planning effort. For our project we will build a replacement panel in lieu of separating the old one to replace damaged stiles and rails. So the statement of work would begin with *remove and replace the damaged*

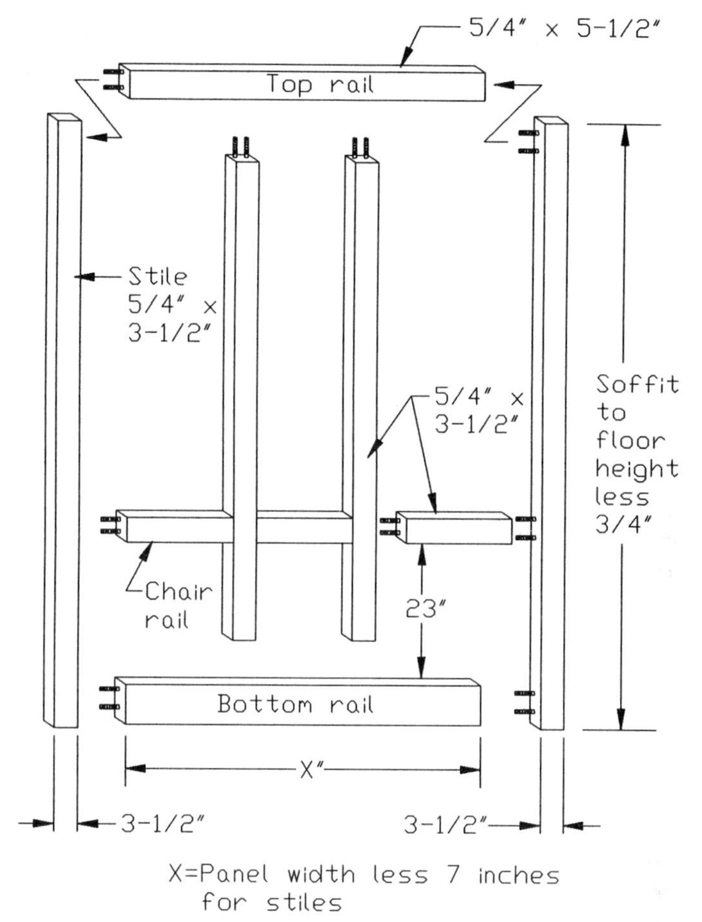

Figure 8-7 Exploded view of panel screen unit.

screen panel. The new one will be made to the same specifications as the old one. We need some further details, such as *the panel will be constructed using either double-dowel or mortise-and-tenon jointing between all stiles and rails. Also, exterior glue will be used.* The finish must also be stated: *Screen wire will match the other panels for type, new moldings will be used, and a three-coat exterior semigloss latex paint system will be applied before the panel is installed.* Finally, we need to add details concerning the reinstallation, such as *the new panel when installed will be raised 1/8 to 1/4 in. above the floor, the quarter-round molding holding the panel in place will be repainted, and new base lattice will be used along the floor and have weep hole notches every 18 in. It too will be painted.*

The planning effort follows other efforts where subcontractors are involved. We need to contact a cabinetmaker or carpenter with cabinetmaking tools that can build a new panel. We also need to contact a painter to do the painting. The cabinetmaker can also be expected to remove the quarter-round and other moldings to remove the panel and take it to the shop. He or she should also be tasked to install the newly built assembly.

Some coordination is required between the subcontractors, but not much. We can handle most of the scheduling from the office.

Contract. We will offer a fixed-price contract to the owner. Our contract with the owner will have minimal costs associated with our company. Except for some administrative and overhead costs, the price will consist mostly of the estimates by the subcontractors.

The body of the contract should be similar to this:

> For the fixed price specified below, we agree to remove and replace the damaged screen panel. The replacement panel shall conform to the measurements and style of the damaged ones. Similar materials of equal or better quality shall be used in the construction. Joinery shall be at the discretion of the cabinetmaker, but will be either double-doweled or mortise-and-tenon type to ensure the quality construction needed. Like the old panel, protection shall be taken to ensure a free flow of water from the porch under the panel. All wood will be primed and double coated with latex exterior paints. The owner shall have the right to select the color.
>
> Total for all work: $×××.××

Material Assessment

Direct materials	Uses/purposes
5/4 × 8	Stiles and rails
Dowels	For use in the double-dowel joints
Exterior glue	Bond joints
Screen wire	Complete the screen panels
Screen molding	Hide the staples and finish the outsides of the panels
Lattice	Base molding
Brads and 4d finish nails	Nail moldings
Exterior latex paint system	Paint the wood
Quarter-round	Used to finish the installation of the panel

Indirect materials	Uses/purposes
Wedges	Temporarily hold up the panel during installation
Plastic sheet	Protect the floor from paint spray
1 × 2 Spruce	Temporary bracing

Support materials	Uses/purposes
Cabinetmaking machines	Build the panel
Bar clamps	Clamp the stiles and rails

Carpenter's tools	Construction
Painter's tools	Apply paint system
Ladder	Install molding
Miter box	Miter molding ends

Outside contractor support	Uses/purposes
Cabinetmaker	Build and install the new panel
Painter	Paint the wood

Activities Planning Chart

Activities	Time line (days) 1	2	3	4	5	6	7
1. Contract preparation	X						
2. Scheduling and materials	X						
3. Remove panel		X					
4. Construct new panel		X	X	X			
5. Install new panel					X		
6. Painting system					X	X	

Reconstruction

Contractor support. The prime contractor will have to negotiate the bidding and have the contract signed. The subcontractors will work through the primary contractor; thus their bids must arrive at the contractor's office in time for the completion of all parts of the bid/contract. Our office personnel will prepare the forms and compile the total bid price.

Materials and scheduling. We have no direct requirements for materials for this contract. Each subcontractor will provide the materials associated with his or her work. The cabinetmaker will remove the old damaged panel and supply a new one. The painter will supply all the paints needed for the job.

We will coordinate the schedule of work with the owner and the subcontractors. The activity chart shows 4 days for the cabinetmaker to perform the work and 2 days for the painter to do his or her work. We also show that the work of the subcontractors can overlap. This may or may not be the actual case, but it is very possible.

Removal of the panel from the porch. The removal of the panel from its surroundings is very straightforward. Working from the inside or outside, the cabinetmaker will remove the quarter-round on each side and across the head. He or she will strip away the base lattice molding and look for the nails that hold the panel against the other molding. When these are found, the cabinetmaker will use a nail set and drive them through or use a wood chisel to expose the heads and pull them out. Once the panel is freed, it can be knocked out by tapping it from the other side.

Building the new panel. Figure 8-7 shows a blowup of the typical construction of a screen panel using cabinetmaking techniques. There are outside and intermediate stiles, a base rail, a head rail, and a chair rail. Notice that the four outside pieces are full length. This forms the frame for the intermediate pieces and adds strength to the unit. Because of this arrangement, a lot of planning must be used to make the pieces fit properly.

The standard 5/4 lumber used to make the frame must be ripped and dressed to correct widths for the stiles and rails. Notice that the side stiles are either made the same width as the middle stiles, or they may be slightly wider to allow for the quarter-round. The rails are all different widths. The top rail is the same width as the outside stile. The bottom rail is almost always $7\frac{1}{4}$ in. wide, and the chair rail can be the same as the head or slightly narrower.

Styling can also be added. A common feature is to make the top rails wider and cut a curve in each.

The assembly of the pieces begins with the center pieces and continues to the outsides. Bar clamps aid in the assembly. Where the mortise-and-tenon joint is used, we can drive cut-off 4-penny nails into the joint after the clamps close it up. Then we can remove the clamps while the glue is still not set and continue with the assembly. If we use double dowels, they will usually hold together after the clamps are removed, even when the glue has not set.

When the entire unit is assembled, we need to surface sand it to smooth out any slight differences that exist at the joints. For this job we use 100- or 120-grit paper in a portable belt sander.

Then we install the wire and screen molding. Our job of building is over.

Reinstall the panel. Reinstalling the panel is simple. We place it in the opening and raise it up with pry bars or wedges until it makes contact with the head or soffit. Then we toenail it in place against the quarter-round. We then cut new quarter-round and install these alongside the posts and across the head. Finally, we cut two pieces of $1\frac{5}{8}$-in. lattice, make notches as inverted V's every 18 in., and nail these pieces flush to the floor.

Painting system. See earlier discussions. A prime coat and two top or finish coats must be applied to all wood.

Concluding comments. The job of either repairing or building a new panel needs to be done by a qualified cabinetmaker or skilled carpenter with cabinetmaking experience. The average homeowner is not equipped to tackle the job. Painting, on the other hand, can be done by many homeowners. This means that the owner can contract the woodworking and reserve the painting for himself or herself.

PROJECT 5. RESTORE THE CEILING AND ELECTRICAL COMPONENTS ATTACHED TO IT

Subcategories include the removal of perimeter molding and plywood panels from the ceiling and replacing them with new ones; correcting problems in the electrical service in the ceiling, including the replacement of a 60-cycle fan; adding new molding around the perimeter; painting after a wash-down to remove mildew.

Primary Discussion with the Owner

Problems facing the owner. The owner can face several problems where the porch ceiling is the area of concern. Some of these relate to the wood ceiling and others relate to the electrical lighting and fans usually attached to the ceiling. Contamination, fungus, mildew, and such readily accumulate on the ceiling and fixtures. On many porches, mildew on the ceiling is a common occurrence. Sometimes moisture and spores unite to form fungus, especially in the crevices and fixtures. Corrosion occurs on metal contacts, wires, and other metal surfaces exposed to the weather even indirectly.

Problems are also encountered with the wood in the ceiling. For example, a ceiling made of 1/4-in.-thick plywood can buckle, warp, sag, and in some cases peel apart if the wrong type was used. Where 1 × 6 V-joint T&G fir or other species wood is used, the owner will frequently request a varnish finish with a slight stain added. This looks great when first finished, but in several years the varnish deteriorates and cracks. Soon the ceiling is unsightly and mildew collects behind the varnish. The exposed stained wood darkens, too.

Problems with a wainscot ceiling are few. Most often the wainscot is painted. The wood is very hard and brittle, but seldom deteriorates. The owner of the porch with wainscot usually has the job of washing away mildew. Since there are many, many grooves in the wood, there are many places where the mildew collects. Painting wainscot is not difficult but is more labor intensive than plywood, for instance, since the paint must be applied to each groove separately with a brush. Sometimes the ceiling can be sprayed. If spraying is done, the painter will usually make a pass through the groove and then, when he or she sprays the flat surface, a second application, although lighter, goes into the groove.

Usually the border of the ceiling is finished with 3/4-in. quarter-round molding or 1 × 2. These present no problems for the carpenter or painter when repairs have to be made.

Alternatives to the problem's solution. To restore the wood parts of the ceiling, we must adopt the position that the finished product will look like new. This premise is very important.

Where the plywood has deteriorated from whatever cause, we must not only replace the pieces, but also attempt to eliminate the cause. This can mean several things. If there is a general sag between ceiling joists or bottom chords in the truss, we must either select a thicker plywood or add nailers at 16 in. o.c.

Another solution possible for the plywood ceiling is to add 2 × 4 blocks every 4 ft and make sure that the plywood is installed with the surface grain perpendicular to the run of the joists. This does two things. First, the edges parallel to the run of the surface grain will have solid nailing, and, second, we are using the greatest strength of the plywood between gaps between the joists.

Where the ceiling is made from V-grooved 1 × 6 T&G, we probably will not need to replace boards. But we will need to strip the ceiling of all finish and replace the finish. For this we will need to burn away the varnish or paint. This is labor intensive because we must clean out the grooves as well. If the ceiling is to be revarnished, we will likely need to bleach the wood and apply a fresh coat of stain before applying the varnish. We will also probably switch from spar varnish to polyurethane.

However, if the wood is damaged, then we will have to find an exact match. This may prove to be a major undertaking. Each batch of V-grooved T&G is supposed to match exactly. Experience tells us otherwise. Usually the thicknesses from lot to lot are uniform, but sometimes they are off by 1/32 in. More often though, the V-groove does not exactly match. It is supposed to be 45° and reach to 1/8 in. of the tongue or groove edge. Sometimes there is slightly less groove or slightly more. The difference is noticeable. The carpenter must make adjustments with a special hand plane called a *rabbet plane.*

Where the ceiling is made from wainscot, the problem of stripping the old paint is more acute than in the V-grooved ceiling. As shown in Figure 8-8, wainscot has half-round curved, molded elements. There is a half-round at one edge and in the center. These moldings give character to the wood. They must be restored to their manufactured depth if the ceiling is to regain its original beauty. Repeated applications of paint obscures the designs. Stripping the old paint from the groove to expose the original wood is labor intensive.

Where the wainscot is damaged from water and needs replacement, we are faced with the same problem as with the V-grooved 1 × 6. The newly milled wainscot may vary slightly from the old. These variations can be in thickness, depth of groove, and width of the bead along the edge and through the center. The only solution for this is to replace the entire run, from end to end. If, for example, six rows are damaged in a corner where the roof leaked, and the ceiling is 16 ft long, we will buy six pieces 16 ft long or the equivalent if bought in shorter lengths. This eliminates the problem of end matching the old with the new.

Electrical wiring and fixtures need our attention due mainly to corrosion. Over time, moisture seeps into the box where the wires from the house service and fixture

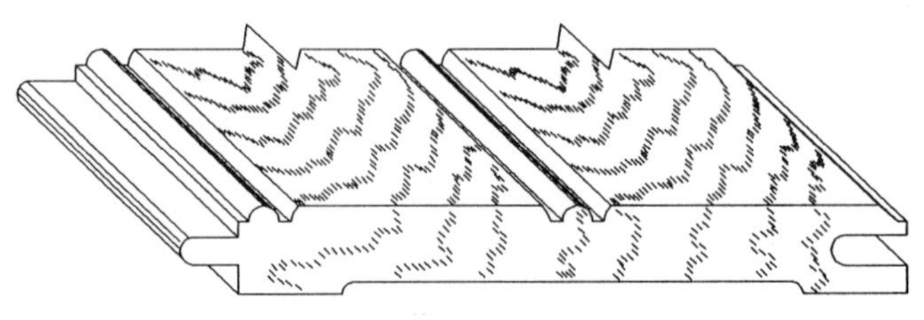

Figure 8-8 Wainscot.

are joined. Since the wires are conductors, they are made from either copper, aluminum, or silver. All three of these metals corrode and oxide, which forms a protective coating. However, the oxidation also form a resistance to the flow of electrical current. When the oxidation is complete, no current will flow from wire to wire and a permanent barrier is formed. The lights and fan will not work. The simple cure is to remove the power and clean all the wires or strip off a small end and wire-nut the shiny ends together.

While inspection is being made, we must also examine the insulation around the wire. Over time this material decays from repeated high temperatures during the summer and very cold temperatures in the winter. If we find cracked or weathered cabling, our only recourse is to replace the entire length. This means that we must climb into the attic and lay a new cable alongside the old one.

Statement of work and the planning effort. For our generic project we will assume that the ceiling is made from 1/4-in.-thick plywood and has been deteriorating for years. It needs replacement. There are two ceiling lights and one has a fan. Our first statement of work is to *remove and replace the entire plywood ceiling.* We must also *add sufficient support materials to prevent sagging between joists.* Since the attic will be exposed, we must *inspect all wiring and connectors and replace any that do not meet modern codes.* The owner desires a pattern in the new ceiling. We recommend a *4 × 4 ft piece of plywood with chamfer edges be installed to create a pattern. Opposite sides will have pieces equal in width.* There must be a finish: *The new ceiling will be finished with a three-coat exterior paint system.* Since we are replacing the plywood, we should also *replace the perimeter molding with 3/4-in. quarter-round.* Finally, *all electrical fixtures and the fan will be inspected for quality and the decision to replace the fixtures will be the owner's. The owner also agrees to purchase the fixtures and have them at the site while work is ongoing.*

The planning effort is not very difficult. We need carpenters to make the repairs to the ceiling. We need an electrician to service the wiring and install the new fixtures, and we need a painter to paint the ceiling. This means that we must provide a coordination plan. We must also obtain bids from the electrician and painter; we are performing the carpentry work. While assessing the job, we determined what corrective efforts were required and how much materials were needed. We also had the electrician make a preliminary visit to the job to determine if the wiring needed replacement.

The office personnel will compile the figures and prepare the bid/contract offer.

Contract. The best contract for this job is a composite fixed-price type. The electrician and painter should have little trouble preparing a solid bid/estimate since they can inspect the job ahead of time and discuss the job with the owner. We can accurately estimate the materials needed to modify the ceiling to prevent sagging, install customized plywood panels, and add new perimeter molding. The body of the contract should state:

> The work of restoring the porch ceiling and fixtures will be done by qualified craftspeople during the time agreed to between the owner and contractor.

The work shall consist of replacing the wood ceiling with a new plywood ceiling customized for the owner. The pattern will be large squares of plywood, beveled at the edges and secured on all four sides to eliminate sag. The end or perimeter pieces shall be equal in width on opposite sides.

A three-coat latex painting system will be applied, and the color will be selected by the owner.

The wiring to the porch will be replaced and new fixtures will be installed. The owner will supply the new fixtures while the job is ongoing.

There will be a complete cleanup after the work is finished.

Total cost: $×××.××

Material Assessment

Direct materials	Uses/purposes
Exterior plywood	Panels for ceiling
1 × 3 Spruce	Lattice work for 4 ft × 4 ft nailing
6d Common nails	Nail 1 × 3's
4d Galvanized finish nails	Nail plywood
3/4-in. Quarter-round	Perimeter trim/molding
6d Galvanized finish nails	Nail quarter-round
2 × 4 Studs	Blocks between joists
8d and 16d Common nails	Nail blocks in place
12-2 with ground wire	Replace the electrical wire
Wire nuts	Make connection in electrical boxes
Exterior latex paint	Prime and top coats for ceiling

Indirect materials	Uses/purposes
Roller replacement	Apply paint to ceiling
Plastic sheets	Cover the floor to prevent spray

Support materials	Uses/purposes
Carpentry tools	Construction
Electrical tools	Construction
Painting tools	Painting
Step ladder	Access to ceiling
Sawhorses and planks	Work surface
Power table saw and cord	Cut plywood into panels
Mason line	Check the level of the ceiling joists

Outside contractor support	Uses/purposes
Electrician	Rewire the service to the porch
Painter	Paint the new ceiling

Activities Planning Chart

Activities	1	2	3	4	5	6	7
1. Contract preparation	X						
2. Scheduling and materials	X						
3. Removing old ceiling		X					
4. Preparation of joists			X				
5. Installing new ceiling			X	X			
6. Rewiring electrical service				X			
7. Painting system					X	X	

(Time line (days))

Reconstruction

Contractor support. As mentioned earlier, this contract involves three trades. As general contractor we will compile the bids from the subcontractors and provide a single contract price. Our office personnel will provide estimates of materials, price them, and add them to the bid price. They will also factor in the overhead for our organization only. We will assume that the subcontractors will also factor in their overhead costs. Our allowance for profit will be our usual mark up plus 1% to 2% for a multicontractor contract preparation.

Materials and scheduling. Each subcontractor will furnish his own materials. As carpenter contractor, we will supply the materials needed for the ceiling and above the ceiling. Since these are all common materials, we will have the two carpenters pick up the supplies on the way to the job site. This transportation cost was figured into overhead.

We will need access to the porch for about 4 days. For two of those days, the carpenters will replace the ceiling. While the attic is exposed, we expect the electrician to run a new cable. But he or she may not want to make two trips to the job. If this is the case, the wiring can be run after the ceiling is up and painted. Then the electrician can do all of his or her work at one time. The painter will need to wait until the carpenters are finished. If the electrician finishes the electrical work before the ceiling is painted, the painter will need to cut in around the fixtures or lower them. This is not a problem.

Removing the old ceiling. On the first day of the job the carpenters will remove the old ceiling. The sequence is as follows:

1. Remove the perimeter molding and take out all the nails.

2. Lower the electrical fixtures, turn off the power, and remove the fixtures. Tape the cable ends separately.

3. Strip the old plywood from the joists.

Preparing the ceiling joists for the new-style ceiling. Figure 8-9 shows a top-down view of the framing needed for the new style 4 × 4 ft panel ceiling. We would use the following schedule:

1. Measure the width and length and determine the width of the side panels. In our example, the ceiling measures 15 ft 6 in. by 18 ft 9 in. Figure 8-9a shows this in detail.

2. With the data from step 1, we mark lines for the 1 × 3's to intersect the joints in the plywood panels. These 1 × 3's are nailed across (perpendicular) to the ceiling joists.

3. Next, we mark the points where the cross members must be installed. One must be installed at the joint of the plywood panels. We snap chalk lines across the 1 × 3's.

4. We use the 2 × 4's and lay them on top of the 1 × 3's at each chalk line unless there is a joist there. With drywall screws, we screw the 1 × 3's to the 2 × 4's. This prepares us for the final pieces of 1 × 3. An alternative is to notch each 2 × 4 to fit over the 1 × 3 so that its bottom surface is flush with the bottom of the 1 × 3.

5. Cut and screw short pieces of 1 × 3 to complete the grid for perimeter nailing of the plywood panels.

This work finishes the formation of the grid. Note that prior to performing the five steps we would have used the chalk line to define any joists that had sagging. Depending on the amount, we may have to make some adjustments. One or two joists are easily raised and trussed to the rafters. If there is a general sag, we may need to create a more extensive truss system. This work and added materials will have been included in the contract.

Installing the new plywood and quarter-round. Since we know the pattern we built for, we can precut all the panels and bevel their edges at one time. For this, we use the table saw and block plane or router and bevel edge bit. The chamfer should be 45° and should be to a depth of 3/16 in.

We can start at any corner or in the middle since we have everything measured, cut, and marked. We use 4d finish nails to nail the panels in place and we set each nail. Nails around the perimeter of the panel will be at 6-in. intervals and through the center at 8-in. intervals.

When we have a panel a fixture must fit through, we mark for the 4½-in. box and cut the hole.

We finish the job by cutting the new quarter-round and nailing it in place with 6d finish nails. Some carpenters use 45° miters. Most use the coping saw method, which makes a finer finished job.

Electrical service repairs. The work of the electrician is straightforward. A new 12-2 cable with ground needs to be run from the circuit-breaker box or fuse box

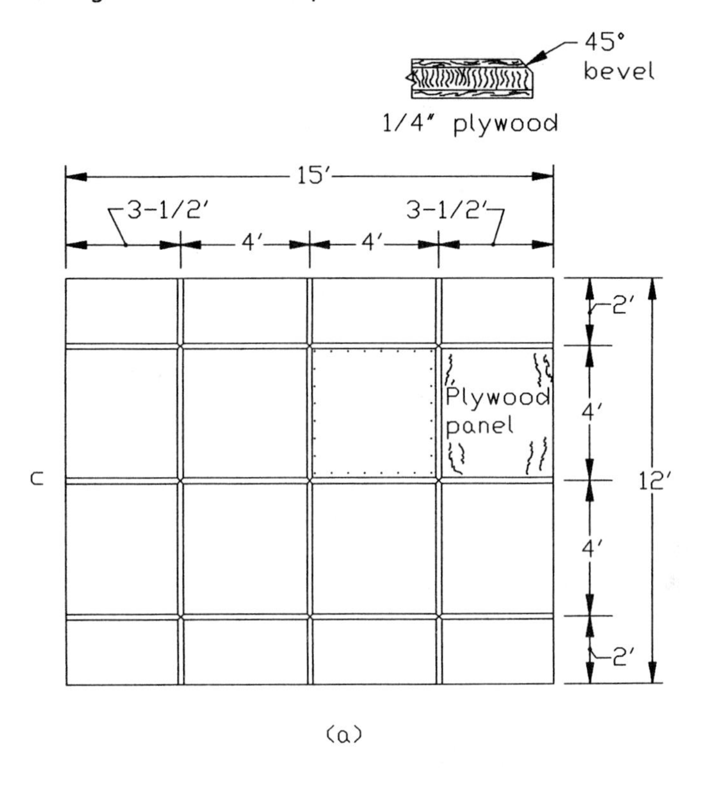

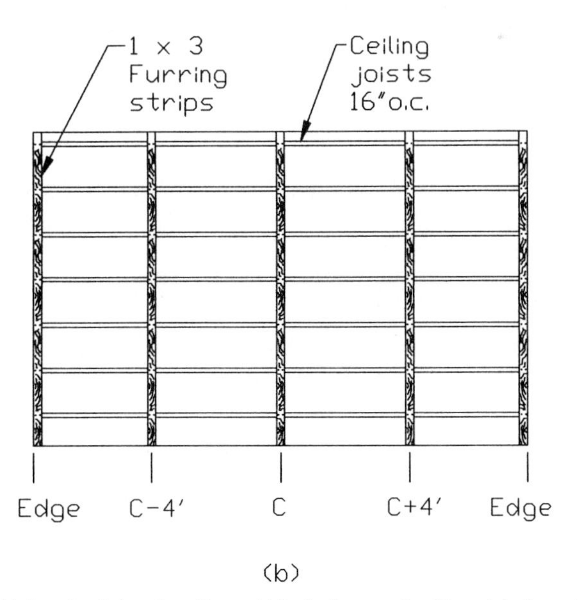

Figure 8-9 Grid showing joists, 1 × 3's, and blocks for panel ceiling: (a) view of panel layout; (b) 1 × 3 layout; (c) 1 × 3 plus 2 × 4 block layout.

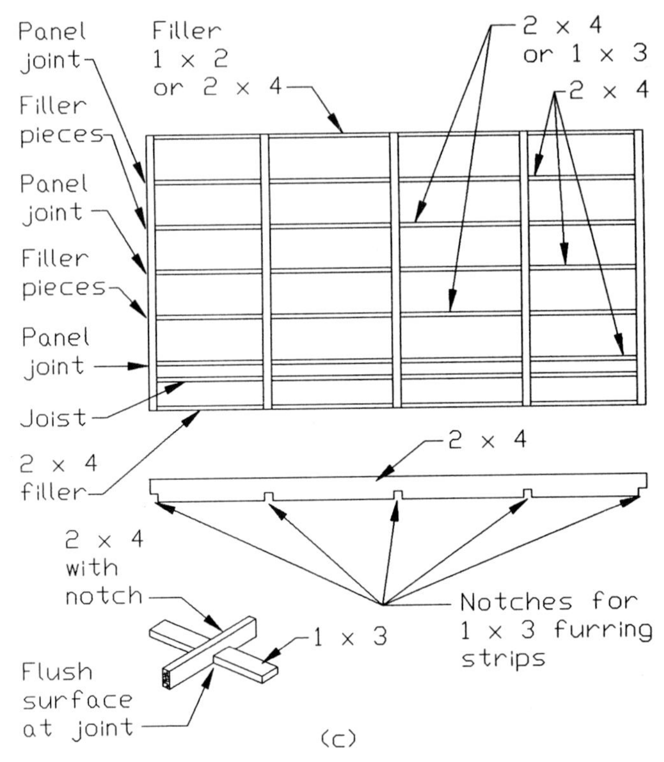

Figure 8-9 *(Continued)*

to the porch ceiling fixtures. Generally, the line runs to a junction box in the attic above the ceiling. From there, short lines run to the wall switch and to the lights.

When these are replaced, the last connection is usually at the circuit-breaker box. This ensures that the electrician works with a dead line.

When replacing the old line, the electrician will usually remove the circuit breaker and physically remove the wires from the service panel first. He or she may actually leave the old wire in the attic, since it can do no harm. A better alternative is to have as much removed as possible to prevent someone else from inadvertently reconnecting it.

The leads from the new lines are stripped about 1/2 in., and these are where the fixtures are connected.

Painting system. We will not discuss the painting system in detail, since we have already discussed exterior latex systems. In our generic project, the owner selects the color. The painter primes and top coats all new wood.

Concluding comments. Restoring the ceiling can be as simple as stripping the paint and repainting. But there are definitely times when the work is more extensive, as in our generic project. We also learned that patterns can be made from

V-grooved 1 × 6 T&G, wainscot, and plywood. These give the porch ceiling character. However, they also create problems when repairs have to be done. New materials frequently do not match accurately. Sometimes we can make small adjustments on the job to blend the new with the old. Sometimes we are forced to replace the entire ceiling to restore it.

It is very possible for the owner to do the work. However, help will be needed since all the work is overhead. It is very difficult to install 1 × 3's, wainscot, or long V-grooved 1 × 6's by oneself. Even the 4-sq.-ft plywood panels are difficult to align and nail alone.

PROJECT 6. RESTORING THE WOOD COLUMN

Subcategories include removing and replacing the base trim and moldings; removing and replacing the crown moldings; restoring the vertical members in a built-up square column; replacing the decayed round column.

Primary Discussion with the Owner

Problem facing the owner. Although this chapter primarily covers screened porches, we recognize that the roof is held up with posts or columns. Sometimes a set of turned posts is used. These are one-piece 5 × 5's turned on a lathe to give style. Other times, columns are made at a mill. These can be round colonial types or square. The core is usually hollow, and the base and crown are added. Very frequently, custom-made columns are built on the site by carpenters. The core of this type is usually a double 2 × 4. The crown and base are custom made, too.

The owner has trouble with posts and columns from rotting and splitting as a rule. Let's discuss each type separately when identifying problems.

The problem with the post type is splitting and rot on the bottom. Generally, the post was cut from the heartwood of the tree. This is the center. Heartwood can be very sound, but it also can have weakened cell structure. The rings are such that they create grain that is subject to splitting. The problem of rot usually occurs at the base. Wood cells weather away, leaving unsightly irregular shapes where before there were fine even cuts. The rot cannot be prevented since the post rests on the floor.

The problem with the factory-made column is usually with the crown and base. Only rarely does the column split. The crown must be designed with a slope away from the house in order to shed water and snow quickly. Sometimes a flat crown piece is used. This creates problems. The base is made from one or two pieces. These are usually molded and are attached to the bottom of the column. It is at the point where the two parts are connected that rot usually starts. Often, the water drains down the column and seeps between the molding and column. It creates rot. At other times the base rots from the bottom where it makes contact with the porch floor. Even though the porch floor is usually sloped, water is trapped on the high side of the column's base and seeps under the base.

In the on-site built-up column, we can have damage from water in the same places, in the crown, on the base, and also throughout the length. There are four side pieces wrapped around a core, as Figure 8-10 shows. The butt joints can separate over time from the action of the weather and sun. Water can leak into the core and cause rot in the base. A poorly designed and built crown is also a major cause of column damage.

Alternatives to the problem's solution. With turned posts, we first fill splits with wood putty and repaint to restore the post to quality. We also sometimes add base molding where there was slight rot to the bottom of the post. If these alternatives are unsuitable, we must replace the post with a new one.

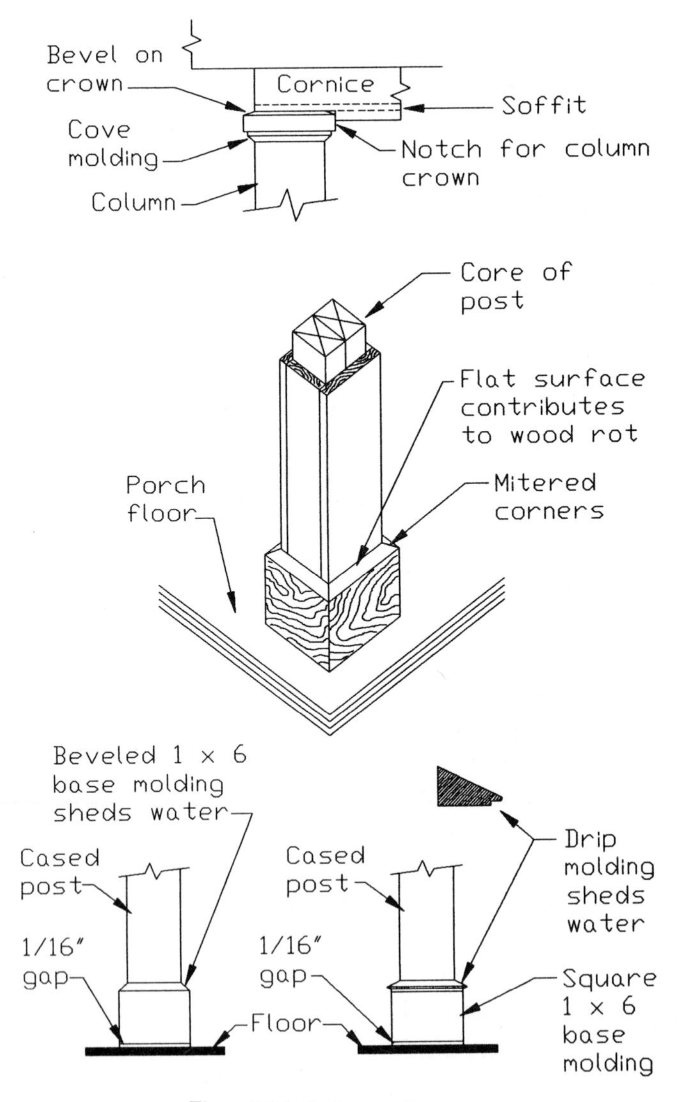

Figure 8-10 Built-up column.

The alternatives to restoring the manufactured round column are to redesign a new crown or order a new one from the factory. The alternative to a damaged base on this type of column is the same as the crown, order a new one from the factory or redesign one and construct it locally. If the column is damaged, we will be hard pressed to make repairs beyond filling small cracks and splits. We might have to order a new one.

The type of column that we have the greatest opportunity to restore is the on-site built-up one. We can strip away the base and crown and even take off the four pieces that form the dimensions of the column. The basic double 2 × 4's remain. When we make repairs, we should improve on the design to make the quality of the column last longer. We can design specific characteristics into the crown to shed water quickly. We can keep the base above the floor by 1/8 in. to permit air to flow under the column and thus prevent rot. We can use treated lumber, which also adds life expectancy.

Statement of work and the planning effort. For our generic project we will restore an on-site built-up column that has extensive damage from weathering and rot. Our essential statement of work is *restore the column to its original quality by replacing any damaged wood and molding.* Since the column was built on site, we need to further state that *base pieces and moldings must match the old ones even if they must be custom made.* We should also state the same for the crown: *Likewise, the crown must match the old one for dimensions and style.* We can even specify the wood to use: *Replacement wood shall be clear stock, fir, hemlock, cyprus, or redwood.* Finally, we must specify the paint system: *After the carpenters are finished, the painters will fill all nail holes and apply a three-coat exterior latex system.*

For this planning effort, we can expect to send one carpenter to the job site. He or she will require electrical tools such as table saw, router, and miter saw. The job must be done by a highly qualified carpenter since this work is classified as finish type. Our plan should also allow the carpenter to make an inspection trip to determine the materials necessary for the restoration. Then, on the day of the job, the carpenter can detour to the builder's supply house or lumberyard and pick out the exact materials required. We need to discuss with the carpenter such things as the expected number of days to complete the restoration, best guess on quantity and types of materials, estimates of the amount of customization required, and times needed for making the trim and molding, and the like.

The painter will need to provide input for the three-coat system. His or her bid will be added to the single contract offer.

Contract. Because of the planning technique specified above, we can offer a fixed-price contract. The body can be as simple as this:

> For the price specified below, we agree to restore the column to its original condition.
>
> Total cost $×××.××

But a better contract body will include several more specifications about the job.

For the contract price of $×××.××, we agree to restore the damaged center column on the screened porch. The exact dimensions, style, and characteristics when restored shall match the other columns. First-quality materials will be used and excellent craftsmanship will be guaranteed.

Some unique trim and molding pieces will be hand machined at the job site to match the original.

The paint system will be a three-coat exterior latex type. The owner will have the option to select the paint or may leave the task to the painter to match the rest of the trim and other columns.

Material Assessment

Direct materials	Uses/purposes
1-in. Stock	Clear stock lumber for the column
2-in. Stock	For crown cap, base, and custom-made molding
4d, 6d, 8d Galvanized finish nails	Nail lumber and moldings
Exterior glue	Bond molding at joints
Latex paint	Prime and top coats

Indirect materials	Uses/purposes
2 × 4's	Replace the basic substructure
Plastic sheets	Ground cover to avoid paint spray

Support materials	Uses/purposes
House jack	To relieve pressure from the roof
Carpentry tools	Construction
Painting tools	Paint the wood
Power machines	Ease work and to make customized trim and moldings
Step ladder	Work surface
Sawhorses and ladders	Workbench

Outside contractor support	Uses/purposes
Painter	Apply the paint system

Activities Planning Chart

Activities	Time line (days)						
	1	2	3	4	5	6	7
1. Contract preparation	×						
2. Scheduling and materials	×						
3. Restoring the column		×	×				
4. Painting the column				×	×		

Reconstruction

Contractor support. Minimal support will be required of the contractor's office staff and estimator. Our project is for a single post; if there are several posts, we simply multiply the amounts for one column to arrive at the total variable costs and add our fixed costs and allowance for profit.

To arrive at a total cost, we will require the painter's estimate or bid price. The contractor will deal with the owner and have the contracts signed.

Materials and scheduling. No unusual materials are needed for this job. All are readily available at well-stocked lumberyards. We know that *select* or *cabinet-grade* woods will be used for the job.

The schedule for contract definition and preparation will take a day to complete after the carpenter makes the first initial inspection and assessment. After contract signing, we program the carpenter for 2 day's work and the painter for 2 day's work. The carpenter will likely spend 2 full days at the job. The painter will not need 2 full days; however, applying multiple coats requires drying time between coats.

Restoration of the column. We have three essential tasks and many specialized tasks. The three essential ones are as follows:

1. Strip away all the old pieces that make up the shape and character of the column. While we do this, we will attempt to retain some without breaking them.
2. Install the shell around the double 2 × 4 basic core.
3. Add the crown and base trims and moldings.

The carpenter will cut four pieces to form the outside square of the column. Two pieces will be 1½ in. narrower than the other two. These will be nailed to the 2 × 4 core and to each other. Then the crown pieces will be custom made and nailed in place. Finally, the base trim and molding will be made and nailed in place, too.

When we build up the base, we will keep all pieces 1/8 in. off the floor to eliminate the water trap and allow some air passage. If any rain water happens to get behind the trim, it will easily flow out.

Painting system. Refer to our previous discussion on applying exterior paints for the primer and top coats. However, in this job we will probably see the painter applying the system with small brushes.

Concluding comments. Columns and posts sometimes need restoration. We have examined the causes for their decay and failure. We have discussed many corrective actions and, finally, we looked at a generic problem and its solution. Some homeowners can perform most if not all the work. But when custom work must be done, a skilled craftsperson must be employed. Joints must fit snugly and accurately.

Special moldings and trims must match the rest of the columns. Quality is a paramount requirement.

CHAPTER SUMMARY

This chapter covers a great many of the problems that can happen when we have a screened porch. We have examined many solutions to each problem. Some we found out can be done by the homeowner who has some carpentry skills. Many, though, need the skills of the tradespeople. All the work is classified as finish. There is no room for mistakes, ill-fit joints, misalignment, and shortcuts.

Appendix

CROSS-REFERENCE TROUBLESHOOTING LIST

This troubleshooting reference listing should make it easy to pinpoint the problems a contractor or homeowner may have with the exterior of the house. First, select the major area of the problem from column 1 (these are in chapter sequence). Then search the middle list to locate a more accurate description of the problem. Then refer to the last column for the project number in the chapter where details associated with the problem are discussed. Even though the subject of the project may be different than the problem, the problem is a subcategory of the project and, therefore, some detail is provided for understanding and solution.

Major topic	Minor problem	Project number
Roof and cornice	Aluminum cornice	3
(Chapter 1)	Close and closed cornices	1
	Close cornice	3
	Damaged flashing	2
	Damaged wall plates and maybe studs	2
	Destroyed wood shingles	2
	Gable end cornices	1
	Painting	2
	Remove the hip rafter	3

Major topic	Minor problem	Project number
	Replace damaged siding	2
	Replace jack rafters	3
	Replace the sheathing and shingles	2
	Restore wood cornice	2
	Splice and replace rafters	2
	Wind-damaged sections	1
	Wind-torn or ice-damaged sections	1
Siding (Chapter 2)	Burning or heat-iron away old paint	1
	Caulking	3
	Corner joint treatments	1
	Double-layered shingles	4
	Hand-split shakes	4
	Insulate behind the siding	3
	Machine-grooved shakes	4
	Naturally aged, painted, and stained shingles	4
	Prestained shingles	4
	Preparation for painting siding	1
	Pressure-washing removal of paint	1
	Random installation patterns	4
	Removal and replacement of reverse board and batten panels	2
	Removal and replacement of bevel siding	1
	Removal and replacement of drop siding	1
	Removal and replacement of 1 × 12 lap siding	1
	Replace aluminum siding	3
	Replace drip caps	2
	Replace vinyl horizontal lap siding	3
	Replace vinyl vertical siding	3
	Sawn shingles, 5-, 6-, 8-, and 10-in. exposure	4
	Spray staining siding	2
Brick veneer, masonry, and stucco (Chapter 3)	Apply wire reinforcement	2
	Clean fungus from the brick	1
	Eliminate the cause for cracked mortar joints	1
	Install expansion joints	2
	Paint stucco	2
	Premixed stucco treatments	2
	Remortar loosened brick	1
	Replaster stucco with cement plaster	2
	Rewaterproof below the ground surfaces	3
	Repair cracks in mortar joints	3
	Repair cracks in stucco	2
	Replace damaged wood under the stucco	2
	Replace decayed blocks	3
	Tuck-point the face stone or Tennessee stone	1
	Tuck-point the mortar joints	1
Exterior wall structures (Chapter 4)	Balloon framing	1
	Built-in pilasters	3
	Damage from dry rot, ants, rodents, fungus	2
	Damage from standing water, dampness	2
	Damaged corner assemblies	1

Major topic	Minor problem	Project number
	Damaged cornice	1
	Damaged joist assemblies	1
	Damaged plates, headers, and sills	1
	Damaged window and door units	1
	Improperly cured lumber	2
	Infected wood installed during the building of the house	2
	Installation of cap block	3
	Installation of rebars, channel blocks, window sills	3
	Lack of ventilation	2
	Out-of-plumb and square walls	1
	Reinforced headers	3
	Reinforced openings (block walls)	3
	Restoration of the basic block wall	3
	Sagging roof line	1
	Stacked block versus running bond	3
	Western or platform framing	1
Doors and windows (Chapter 5)	Add new types of weatherstrip	2
	Broken door stops	3
	Broken or inoperable locks	3
	Broken panes of glass	3
	Damaged door surface	3
	Damaged or destroyed weatherstripping	3
	Decayed exterior door trim	3
	Defective or missing striking plate	3
	Freeing paint-shut sash	1
	Ill-fitted door	3
	Reglaze the loose glass panes	1
	Refit the stops to restrict sash travel	2
	Refit window stops	1
	Replacement of sills and aprons	1
	Replacement of stops	1
	Replace awning controls	1
	Replace cylinder sash controls	1
	Window and door units	
	Replace sash cords or chains	1
	Replace the bronze strips or other insulation	1
	Replace the composition weatherstripping pieces	2
	Replace the metal weatherstripping strips	2
	Separated stiles and rails	3
	Split or cracked jamb	3
	Stripped screw holes in door or frame	3
	Strip sash of paint	1
	Use filler strips to reduce drafts and loss of heating and air conditioning	2
	Warped door	3
	Worn or rotted threshold or sill	3
Wood decks, porches, stairs, and patios (Chapter 6)	Fitting deck and flooring under columns and around posts	1
	Fit pieces of tongue-and-groove between good ones	1
	Nail flooring with face nails and toenailing	1

Major topic	Minor problem	Project number
	Paint pressure-treated and nontreated materials	4
	Precoating treated materials	4
	Presecure the railings, seat backs, seats, seat supports, and newel posts	2
	Remove the stain from the deck	4
	Replace parts of the railings, stairs, and treads	2
	Replace the 1 × 4 flooring on a porch	1
	Replace the deck on an aboveground patio and deck	1
	Replace the stair treads	1
	Replace defective joists, girders, headers, cross-bridging, sills, posts, and flashing	3
	Replace the 1 × 4 flooring on a porch	1
	Replace an aboveground deck	1
	Replace the stair treads	1
	Treat ground and building materials, footings and piers, stair stringers, rafters on covered decks and porches	3
	Treat pressure-treated materials	4
	Trim overhangs	1
	Use joist hangers and other metal fasteners	3
Concrete slabs, sidewalks, and driveways (Chapter 7)	Add control and expansion joints	3
	Float and finish with a steel trowel	2
	Form the perimeter	2
	Form the sidewalk	1
	Estimate the amount of concrete required	1
	Patch with the intent to cover the driveway with blacktop	3
	Pour and finish the concrete	3
	Prepare the subsoil	1
	Select and create the finished appearance	1
	Select the proper climatic conditions for pouring and finishing	1
	Use of wire mesh	3
	Use a footing–slab combination pour	2
	Use rebars and wire mesh	2
Wood porches with screened panels and wood columns (Chapter 8)	Add new molding around the perimeter	5
	Application of new wire and molding	4
	Correct problems in the electrical wiring and fan	5
	Fit and hang the new door (wood and metal)	2
	Fit the stops	2
	Install a new lock unit	2
	Install the door control assembly	2
	Install the new screen wire	3
	Nail the molding over the staples	3
	Paint after a wash-down to remove mildew	5
	Paint the top coat	1
	Paint the wood door	2
	Prepaint the wood	4
	Prepare and install new treated members or naturally resistant wood members	3

Major topic	Minor problem	Project number
	Prime the new molding	1
	Reinstallation of the panel	4
	Removal of perimeter molding	5
	Remove and replace a round column	6
	Remove and replace the screen wire or cloth	1
	Remove and replace plywood panels from the ceiling	5
	Remove the molding and old wire	3
	Remove the old damaged panel	4
	Remove the quarter-round molding, lattice, or base boards	4
	Remove the screen molding	1
	Repaint the new wood	3
	Repair or manufacture a new panel	4
	Replace base moldings on columns	6
	Replace crown moldings on columns	6
	Replace molding after the wire is changed	1
	Restore columns	6
	Restore the wainscot ceiling	5
	Stretch the wire as it is installed	1
	Take out the rotted vertical and horizontal panel members	3

INDEX

E

F

G

Q

R

T

U

V

W